T0294366

Pecan

Pecan

America's Native Nut Tree

LENNY WELLS

The University of Alabama Press   •   Tuscaloosa

The University of Alabama Press
Tuscaloosa, Alabama 35487–0380
uapress.ua.edu

Inquiries about reproducing material from this work should be
addressed to the University of Alabama Press.

Typeface: Adobe Caslon Pro

Front cover image: Shutterstock, three pecan trees © skizer
Spine images: iStockphoto, watercolor clip art illustrations
of nuts and seeds © Shlapak Liliya
Back cover image: Pecans, courtesy of Kimberly Hatchett
Cover and interior design: Michele Myatt Quinn

Excerpt from SILENT SPRING by Rachel Carson.
Copyright © 1962 by Rachel L. Carson, renewed 1990 by Roger
Christie. Reprinted by permission of Houghton Mifflin Harcourt
Publishing Company. All rights reserved.

Publication made possible in part by generous contributions from
The Samuel Roberts Noble Foundation, Mason Pecans Group,
and Ellis Brothers Pecans, Inc.

Library of Congress Cataloging-in-Publication Data

Names: Wells, Lenny, 1971– author.
Title: Pecan : America's native nut tree / Lenny Wells.
Description: Tuscaloosa : The University of Alabama Press,
[2017] | Includes bibliographical references and index.
Identifiers: LCCN 2016022152| ISBN 9780817318871 (cloth : alk.
paper) | ISBN 9780817388966 (e book)
Subjects: LCSH: Pecan. | Pecan industry.
Classification: LCC SB401.P4 W45 2017 | DDC 634/.52—dc23
LC record available at https://lccn.loc.gov/2016022152

For my mother, Lenn, who gave me the love of words,
and for my grandparents, Marvin and Nita Wells,
who gave me opportunity.

If you would know strength and patience, welcome the company of trees.

—Hal Borland

Contents

List of Figures   xi

Preface   xiii

Acknowledgments   xv

Introduction   xvii

• 1 •
Origins of the Pecan   1

• 2 •
The Legacy of Antoine   31

• 3 •
The Secret Life of Pecan Trees   61

• 4 •
The Rise of an Industry   99

• 5 •
A Tree without Borders   173

• 6 •
Healing the Land with Orchards   195

• 7 •
Rebirth   213

Notes   237

Index   257

Recipes to follow page   224

Figures

1. Centennial pecan trees photographed in 1941   41

2. Colonel W. R. Stuart, "Father of the Paper-Shell Pecan"   48

3. Pamphlet from the Keystone Pecan Company   106

4. Early-twentieth-century pecan orchard   108

5. Spraying pecan trees for pecan scab in the 1930s   117

6. Dusting pecan trees in the 1930s   117

7. The J. H. Burkett family, namesake of the Burkett pecan   132

8. Fabian Garcia, father of the New Mexican food industry   153

Color insert to follow page   171

Preface

In late summer 1528, after plundering through the Florida wilderness, a group of ill-fated Spanish explorers led by the veteran conquistador Pánfilo de Narváez set sail into the Gulf of Mexico from a shallow bay somewhere near the Wakulla River. The expedition had already undergone extreme hardship. Leaving Cuba in February, they were immediately directed off course by an incompetent pilot named Miruelo. Bound for the Rio de las Palmas, today known as Mexico's Rio Soto la Marina, located about 125 miles south of the Rio Grande, the party limped ashore in April near Tampa Bay, believing the Rio de las Palmas lay just to the north. Narváez divided the expedition in two, sending one group overland while crew members and women traveled aboard the ships, planning to meet again at the Rio de las Palmas. On the overland journey, the 300 men became lost. Some were killed by Indians and others became sick and died of disease. Desperate to leave Florida, the men melted their spurs, stirrups, and breastplates in an attempt to construct axes and saws with which to build rafts. They killed their horses to avoid starvation and lashed the logs of their rafts together with the hair of their mounts. Five rafts left the shallow inlet they named the "Bay of Horses" after the animals that were slaughtered there. Following the coastline, they became separated over the next few months and each one landed on a different part of the Texas coast.

Over the course of the next six years, those who survived spent their harrowing days enslaved by Indians. Most were killed. Ultimately, only four men would survive to meet up once again with European settlements in Mexico City. These castaways were the first outsiders to see the interior North American continent. They traveled from Florida to Texas and deep into Mexico to the Sierra Madre, observing plants, animals, landscapes, and people that no European had yet seen.

One of these men, Álvar Nunez Cabeza de Vaca, the treasurer of the original expedition, left a written record of his experiences. His account provides the first known description of an edible tree nut that we call the pecan. The fall of 1532 found these survivors encamped with the Yguase, Mariame, and Quevene Indians along the "river of nuts," now known as the Guadalupe River. To the castaways, these nuts must have been manna from heaven. The trees were loaded with a large crop of pecans and the nuts rained down from their great overhanging canopies to provide nutritious, filling, golden kernels inside thin shells. The Indians came from as far as 90 miles away to feast on the nuts in the fall. There they would gorge themselves on nuts throughout the winter, sometimes almost exclusively. The survivors of the Narváez expedition were undoubtedly the first Europeans and African to taste the sweet flavor of the pecan.

Since that time, pecans have been cultivated in at least 10 countries on six continents. Although consumed by humans for centuries, pecans were not domesticated until the late nineteenth century. The pecan industry as it is known today was developed several years after the dawning of the twentieth century. Thus, in a relatively short period of time the commercial pecan industry arose from obscurity out of the wilds of North America.

Acknowledgments

Many thanks are expressed to all who have opened the door for me into the world of pecans. These include Tom Crocker, Mike Smith, Darrell Sparks, Patrick Conner, Bruce Wood, Hilton Segler, and Bill Goff, each of whom has provided excellent tutelage at some point in my career. I am deeply indebted to the pecan growers of Georgia, who continue to help me learn more about the pecan each day. I am particularly grateful to the pecan growers in and around Albany, Georgia, who were invaluable to my early pecan training when I served as their county extension agent.

I owe thanks to my father, Frankie Wells, who believed enough in my capabilities to partner with me 12 years ago in growing pecans and planting orchards on our family farm. How I wish he were here to see them bear fruit. The experiences in those orchards have helped me understand pecans from a practical perspective and have made me better at what I do.

Thanks to the heirs of J. B. Wight, whose donation of a cardboard box full of early pecan literature to my predecessor, Tom Crocker (who later passed it down to me), became the impetus for this book. Jean Richardson Flack's excellent PhD dissertation, *The Spread and Domestication of the Pecan in the U.S.* (University of Wisconsin), was a very valuable source of information on the domestication of the pecan and its early use in America, as was Jane Manaster's *Pecans: The Story in a Nutshell* (Texas Tech University Press, 2008), particularly regarding the pecan's history in Texas.

Thanks to my editor at the University of Alabama Press, Elizabeth Motherwell, who provided patient and valuable guidance to this first-time author throughout the publishing process.

I am deeply grateful to my wife, Paige, for putting up with the many hours I spend in pecan orchards or with my nose buried in books. Thank you also to Dana and Molly, my little girls, for their patience with their Daddy, and for the joy they give me each day.

Introduction

Centuries ago, somewhere in southern Texas or perhaps farther south into what is now Mexico, a towering tree grew in the bottomlands along a riverbank. Furrowed ridges of grayish-brown bark ran up the length of the tree's trunk and along its branches, disappearing into a canopy of leaves among which grew football-shaped nuts concealed within a lime-green husk. It was late fall and the tree's leaves had turned from a deep, dark green to yellow. Many were beginning to drop to the loamy soil below. As the nights grew cooler, the husks began to open as if called by some unseen force, and the nuts dropped to the ground below with a soft, dull thud.

In the sun-dappled forest, surrounded by other trees and undergrowth, a man reached down to pick up one of the nuts from the forest floor. He had eaten nuts before—bitter-tasting acorns, hard-shelled hickories—but this nut was different. Its shell was somewhat thinner and easier to crack open. Its taste was sweeter, more subtle than the sharp-flavored nuts he had tried earlier. As he began to look around, he noticed other trees like this one. Their nuts were similar, yet with slight differences—some more rounded, some harder to crack open, some filled with less nutmeat. He began gathering the nuts, selecting the best-tasting ones from under their respective trees and placing them in a fur-lined pack for his journey upriver.

Reaching his destination, the man showed the nuts to other members of his tribe. He took them to the nut-filled bottomlands so that they could collect nuts as well. They exchanged the nuts in trade and began to plant them near favored campgrounds on their nomadic journeys each year. Over time, some of the planted nuts grew into trees and later bore nuts of their own on which the tribe would feast through the winters.

Several thousand years later, it's Thanksgiving. Grandmothers across the United States are serving turkey, dressing, cranberry sauce, giblet gravy, and biscuits so good they place these culinary matriarchs on a pedestal. Cooling on the dessert table is a moist, flavor-packed pie sweet enough to keep dentists in business for years to come. Much of the pie's flavor and crunch results from the ingredient for which it is named, the pecan. But outside of its use as an ingredient in this pie, pecans exist in a cloud of mystery to the average person. Unless you are from the areas in which the pecan grows, there's a good chance you don't even realize pecans grow on trees. And what a storied tree it is.

There's more to growing pecans than meets the eye. In the old days of pecan production, the trees' crops were usually grown with the minimal input of only a little fertilizer. At harvest time, the nuts were beaten from the trees with long bamboo poles and picked up by hand. Today, pecans are grown in many ways, from the minimalist approach of those with a tree or two in their yard to methods of varying intensity applied by commercial producers. Geography, environment, and pecan variety largely determine the practices employed by the commercial pecan grower.

In the humid southeastern United States, conventional pecan orchards may be sprayed with fungicides an average of 6 to 12 times to ward off pecan scab disease from the time the year's new leaves pop out in April until the shell of the nut hardens inside the shuck in mid-August. Throughout the growing season, the trees are fertilized, checked for insects (which are sprayed as needed), and irrigated at varying amounts as the crop's development dictates, using little sections of plastic tubing that drip water into the trees' root zone or spread it with miniature sprinklers beneath the trees' canopy as a fine mist. Grasses and competing weeds are eliminated with herbicide along a 12-foot strip centered along the row of trees, while the remaining 70 percent of the orchard is covered with a varying mix of grasses and legumes.

In the arid western production region, the growth and form of the trees are shaped by "hedgers." These large, lethal-looking, $200,000 machines have a long arm tipped by rotating 24-inch circular saw blades capable of pruning off large limbs, branches, and twigs to keep the trees at the artificially limited height of less than 40 feet in a process called "hedging." This allows pecan producers to maintain closer tree spacing in the orchard, increasing the number of trees per acre and increasing yield in the process. Western producers typically don't have to spray for pecan scab since the climate is too dry for development of the disease. Insect pests tend to follow the trees wherever they are planted, however, and must be sprayed when their populations get out of control. Fertilizers are liberally applied to boost production, and the tabletop-level orchards are kept free of other vegetation. The trees are irrigated by periodic flooding or, increasingly in this era of water shortage, with more efficient drip or microsprinkler irrigation.

As fall arrives, the nuts mature, the shucks begin to split, and the pecans are vibrated from the tree under the great force generated by cumbersome-looking, but surprisingly nimble, three-wheeled shakers. These bear monstrous, hydraulically powered pinchers that grip the limbs and trunks between rubber pads. Depending on the tree's size and the nuts' maturity, less than 10 seconds of vibration from these machines can be enough to drop every nut from the tree.

Once the sunshine dries the nuts on the ground for a day or two, giant cylindrical brooms called sweepers are mounted to the front of tractors and large, enclosed fans called blowers are mounted to the rear. The pecans are then brushed and blown along with leaves and sticks into neat windrows. The mechanical harvester then comes along sweeping and sucking the material into its bowels, where the nuts are separated from the sticks, leaves, and other debris by more powerful fans, chains, and belts and dumped into a buggy following behind. When the buggy is filled, the nuts are dumped into a wagon and taken to the cleaning plant for further separation, sorting, and sizing, finally to drop by the ton into oversized poly bags at the end, where they are stacked two high in the warehouse until sale or shipment. If it sounds more complicated than you thought, keep in mind that this is a highly simplified overview. We'll get into the details later.

Through the ages, humankind has developed its own favorite uses for the pecan. The Native Americans had powcohiccora, a strong ceremonial drink made from crushed hickory nuts, including the pecan. Modern Americans have pecan pie and pralines. As ingrained as pecan pie has become in Southern culture, and to some extent in American culture, it would seem that this treat has been around since Colonial times, or at the very least, that it is a relic of the Old South. However, the real story of pecan pie has a more recent origin. Oddly, one of the most reliable methods food historians use to verify a confectionary origin is to peruse old cookbooks and newspaper recipes.

Doing so has revealed that several versions of pecan pie have been around since at least 1886, when a Texas admirer wrote, "Pecan pie is not only delicious, but is capable of being made a real state pie. The pecans must be very carefully hulled, and the meat thoroughly freed from any bark or husk. When ready, throw the nuts into boiling milk, and let them boil while you are preparing a rich custard. Have your pie plates lined with a good pastry, and when the custard is ready, strain the milk from the nuts and add them to the custard. A meringue may be added, if liked, but very careful baking is necessary." Although not necessarily the pecan pie as we know it today, this is the first record of using pecans in a pie recipe.[1]

In the early 1970s, historian Sue Murch was interested in life in Texas during the early 1900s, specifically in what people ate and how it was prepared. As part of her research, Murch interviewed a number of ladies in Dallas-area nursing homes. Out of this she developed a cookbook that contained a portion of the interviews and recipes she obtained during her research. Among the ladies Murch spoke with, one of them, Mrs. Vesta Harrison, claimed to have developed the original recipe for Texas pecan pie.

Mrs. Harrison moved to Fort Worth with her family at the age of four. When she was 17 she saw an ad in the newspaper announcing a cooking school to be held in Fort Worth by a Mrs. Chitwood of Chicago. On the night before attending the school, a recipe came to Mrs. Harrison in a dream. The next day she asked Mrs. Chitwood if she had ever made a pecan pie, to which Mrs. Chitwood replied, "There is no such thing!" "Well," said Mrs. Harrison, "by gollies, I don't know how, but I'm gonna mess up something making a pecan pie!"

According to Mrs. Harrison, "I just made a sorghum syrup pie and put a cup of pecans in it, and it was good." Upon tasting Mrs. Harrison's pie, Mrs. Chitwood was so impressed that she sent the recipe to Washington for entry into a contest. A few days later, they were informed that Mrs. Harrison's Texas Pecan Pie had won the $500 prize.[2]

While Mrs. Harrison may or may not have come up with the original version of pecan pie, her experience would have occurred about 1913. Thus, her recipe could very well have been one of the first, if not the first, for something resembling pecan pie as we know it today—or at least, she could have believed so. This was not exactly the "information age," and news probably didn't travel all that fast. Even if there were already a pecan pie recipe in existence, it would probably have still been largely unknown.

While early recipes often called for sorghum and molasses, the pecan pie as it is generally known today was born following the invention of Karo corn syrup by the Corn Products Company in 1902. One of the earliest recipes for pecan pie using Karo syrup was penned by Mrs. Frank Herring in the Sallisaw, Oklahoma, *Democrat American* on February 19, 1931: "3 eggs, 1 cup Karo (blue label), 4 tablespoons corn meal, ½ cup sugar, ½ cup chopped pecans or less if desired, 2 tablespoons melted butter, pastry. Method: Beat whole eggs slightly, add Karo, corn meal, sugar and melted butter, then stir all thoroughly. Line pie tin with flaky pastry and fill generously with mixture. Sprinkle chopped pecans on top, bake pie in a moderate oven until well set when slightly shaken."[3]

One of the most colorful descriptions of pecan pie comes from Marjorie Kinnan Rawlings, author of *The Yearling* and *Cross Creek*, who considered true Southern pecan pie "one of the richest, most deadly desserts known" and "a favorite with folk who have a sweet tooth, and fat men in particular." Rawlings described two pecan pie recipes, "My Reasonable Pecan Pie" and "Utterly Deadly Southern Pecan Pie," the last of which she declared she served only to "those in whose welfare I took no interest." So rich was this last recipe,

Rawlings claimed to never eat a full portion because of her self-acknowledged inclination toward plumpness and a desire to see out her days on earth.[4]

One pecan pie historian, Edgar Rose, was convinced that chopped pecans were the key to a good pie. His interest in the pecan pie was piqued when he made several trips to Indiana in the 1970s and became intrigued not only with the neat rows of native pecan trees scattered around the town, but with the pie itself. In Rose's opinion, chopped pecans were clearly superior to whole pecan halves for pie making: "Pecan halves may look prettier but they don't get as crisp." He also shunned the popular Karo syrup used in many Southern recipes, preferring instead brown sugar. Rose suggested that Southerners tended to prefer the sweeter taste of a pecan pie made with Karo. In fact, one Southern taste-tester told Rose, "The pie has to be so sweet it makes your teeth hurt."[5]

Like pecan pie, the praline has a rather cloudy past, though a much older one. There are a number of stories surrounding the creation of pralines, but they all center on one man, Clement Lassagne, personal chef of the seventeenth-century French diplomat Marechal du Plessis, Duke of Choiseul-Praslin, for whom the dessert is named. Lassagne actually made his first pralines with almonds, which were coated with grained, caramelized sugar.

Several legendary versions of the exact origin of Lassagne's creation exist. One suggests that upon seeing a kitchen boy nibbling at leftover caramel and almonds, Lassagne had the idea of cooking whole almonds in sugar. Another suggests that the chef discovered his own children caramelizing almonds stolen from his kitchen, being led to the thieves by the aroma emanating from their hideout. A third story tells of a Lassagne apprentice clumsily dropping some almonds into caramel made with honey. Yet another story paints the duke himself as something of a ladies' man who asked Lassagne to develop a deliciously sweet treat that he could provide to his companions. Whatever the true origin of the original praline may be, Lassagne perfected the recipe and eventually retired to Montargis and founded his own confectionary shop, Maison de la Praline, which still exists today.

The close ties of the city of New Orleans to the history of pecans occupy a portion of the praline story as well. As mentioned earlier, the first pralines used almonds as the nut of choice. As the French immigrated to the New World, they brought along their culinary tastes. The appearance of the praline in America is tied to Ursuline nuns sent to the French colony of Louisiana in 1727 as chaperones for the young women known as casket girls, who, at the request of Bienville, came over as brides in training for the male colonists of New Orleans.

As unseemly as this may sound, such activity was not uncommon and certainly, in the case of the casket girls, was not looked down upon. Unlike the female prisoners and street walkers shipped over to many colonies, the casket girls were handpicked from church orphanages and were said to be conspicuous by their virtue. The name "casket girls" was derived from the fact that they arrived in the French settlement with all their worldly possessions crammed into casket-like boxes. The nuns were to instruct the girls in the development of good character, scholastics, and domestic skills. Among these skills, the art of making pralines was included.

As the girls were married off and settled throughout New Orleans and southern Louisiana, their early foundations led to the development of the Creole culture and the spread of pecan pralines. Since almond trees were in short supply in colonial Louisiana, the ubiquitous pecans were substituted into the creamy, sweet, nut-laden recipe. The treats soon became a fixture in the fledgling society of New Orleans.

By the late nineteenth century, free women of color were selling pralines on the streets of the French Quarter. These "pralinieres," as they came to be called, were usually found along Canal Street, near Bourbon and Royal Streets, or around the entrance to Jackson Square. This activity provided these women with a good source of income and the ability to provide for themselves and their families without having to become indentured servants during a trying period.[6]

As interesting as the stories of pecan pie and pralines may be, they only scratch the surface of the pecan's saga. With such history, it is somewhat remarkable that, as foods go, the pecan remains so obscure to so many. Even across the Southern tier of states, a region that reveres its history, pecans are known mainly as that nut that you eat in pies during the holiday season, or that you may see sweetened and sold as snacks in candy shops, or salted and roasted, or sometimes even raw in the shell at roadside stands. Even here, among its "home" states, the story of the pecan and its place in our culture remains largely untold. That story goes much deeper than pecan pies and pralines. How did this football-shaped nut found by tribespeople those thousands of years ago make its way to your grandmother's table? How might we benefit from allowing pecans to take on a larger role in our diet rather than simply using them as toppings and ingredients for holiday treats?

I was once in a meeting in which the ever-emerging positive effects of eating pecans were being discussed, and someone made the statement, half in jest, that we should call the pecan the "tree of life." A number of other plants, most notably the arborvitae, are known as the tree of life. The arborvitae earned its nickname because during the seventeenth century, Europeans believed that its sap held healing powers, a false belief disproven by modern science. In fact, if swallowed, the sap of the arborvitae is toxic and can cause serious side effects. In contrast, my colleague's reference to the pecan as the "tree of life" is not so far-fetched. Based on Cabeza de Vaca's description of the reliance of certain Native American tribes on pecan nuts throughout a portion of the year, the pecan tree truly was a "tree of life" for these people. While they may not have known that the small nuts they were eating were packed with protein, fiber, healthy fats, nutrients, vitamins, and antioxidants, de Vaca and the Native Americans knew firsthand the nutritional value of the pecan.

The life-giving traits of the pecan tree do not end with the nutritive qualities of its fruit. There is a popular bumper sticker that reads "Trees are the Answer." Indeed, as we shall see, trees, including pecan, are one answer to many of our world's problems. J. Russell Smith championed an agriculture based on trees in his classic book *Tree Crops: A Permanent Agriculture*, extolling the virtues of tree cropping systems to protect the soil while producing healthy and valuable agricultural products. In addition to these virtues, we know that properly managed orchards can store carbon, prevent air and water pollution, and serve as habitat for myriad species.

The real story of the pecan is not only that of the nut itself but that of the tree on which it grows, and indeed that of the men and women throughout recorded history who have contributed to its domestication and development as a commercial crop that is now spreading across the world. I've spent my life in southern Georgia, where pecan trees are everywhere and where more of their nuts are produced than anywhere else in the world. Yet I had never heard the real story of the pecan until I began piecing it together from a box of obscure old papers, pamphlets, and articles left by my predecessor when I became the "pecan specialist" at the University of Georgia.

Within that box I found old magazines from as far back as 1902 with titles like *The American Nut Journal*, pamphlets from the early 1920s printed by the long-defunct Keystone Pecan Company, and letters of correspondence between pioneering agricultural scientists and gentleman farmers of a bygone era. As I began to read through this old, crumbling material, it dawned on me that I was reading not only about the domestication of a crop in relatively recent times and the birth of what we know today as a thriving agricultural industry in the southern half of the United States, but also about how intimately the pecan is woven into the fabric of our nation's history.

Pecan

· I ·

Origins of the Pecan

We all travel the Milky Way together, trees and men.

—John Muir

FROM THE FORESTED BOTTOMLANDS of the mighty Mississippi River and the rivers of southeastern Texas and Mexico, pecans have emerged to become a valuable crop in the southern United States. Very few food crops offer a detailed record concerning their cultivation and spread as an agricultural industry. Many have been utilized for so long that records either do not exist or have been lost to the ages. For many crops, such as wheat and rice, the wild ancestral species from which the crop was derived no longer even exist.

The domestication of perennial tree crops is a much slower process than that of annual crops because of the extended period of time required for tree crops to bear fruit. Most fruit and nut crops introduced to North America underwent many generations of human selection before being brought to the New World. The peach, for example, has been cultivated in China since at least 1100 BC and finally found its way to America in the seventeenth century, when George Minifie planted peach trees at his estate in Buckland, Virginia.[1] The domestication of the pecan, by comparison, is still in its infancy. Because of its recent introduction to cultivation, we have a relatively good record of its domestication, offering an interesting glimpse of this species as it developed into the crop we know today.

Domestication can be an ambiguous term, depending on the context in which it is used. From the botanical or evolutionary perspective, pecans are still considered to be relatively "un-domesticated" because of the genetic similarity between cultivated and "wild" pecans. From the layman's perspective, the pecan is domesticated in that it has been adapted to life in association with and to the advantage of humans. Using this working definition, a species is maintained through isolation of the plant from its wild population and the selection of traits based on a certain set of requirements. In the case of the pecan, these traits may include nut size and quality, disease and insect resistance, precocity or earliness of bearing, and harvest date, among other characteristics. In most cases, human selection runs counter to natural selection in order to shape the plant's traits for human use. As we will see, pecan production provides a good example of how we have adapted our growing techniques to meet the genetic capabilities of the tree without significant alteration of the tree's genetic makeup.

The story of the pecan, in many ways, reflects that of our own nation. The humble nut is intertwined with the path of American history from the earliest inhabitants of the North American continent to the conquistadors, our founding fathers, the Old West, the New South, and the global economy. Rodney True, one of our foremost agricultural historians, called the pecan "America's most important contribution to the world's stock of edible nuts."[2]

Since Cabeza de Vaca's description of the pecan following his long and arduous journey, pecans have been linked to the history of the New World. Indians and settlers alike utilized the pecan as a food source and trade item before the nation came into being and throughout its infancy. As the nation grew and developed, so did the spread, utilization, and eventually the cultivation of pecans into a thriving industry in the southern half of the United States. From there, pecans have spread to each continent of the world, with the exception of Antarctica. Yet despite its growing international recognition, the pecan is still unknown in many parts of the world.

A Pecan by Any Other Name

The pecan is scientifically classified as a hickory, belonging to the walnut family (Juglandaceae). The word "carya" was coined by biologist Thomas Nuttall in 1818 to differentiate hickory trees from more distantly related walnut trees; it is derived from the Greek *karya*, which means "nut tree." The word "pecan" has many pronunciations, all of which originate from the Native American word *pakan*, meaning nuts that require a stone to crack or "a hard-shelled nut." To many Native Americans, *pakan* referred not only to the nuts that we know today as pecans, but also to hickories and walnuts.[3]

The history of this word survives by rather inauspicious circumstances. Around nine in the morning on November 28, 1729, a group of Natchez Indians attacked Fort Rosalie, a French garrison made up of small farmers and planters in what is now the city of Natchez in Adams County, Mississippi. Established in 1716, the post served to protect a settlement established two years prior from the Indians. The Natchez were considered to be among the most civilized of all the original inhabitants of North America. Upon their arrival to the area, the French were greeted by Natchez warriors bearing their king, who was called "The Great Sun," on their shoulders. The Natchez king welcomed the Frenchmen and a treaty of friendship was struck, in which the Natchez allowed the French to establish the fort and a trading post.

The friendship between the French and the Natchez people was betrayed by a single man, M. de Chopart, commander of Fort Rosalie. The commander was harsh in his treatment of the friendly tribe, finally ordering the Great Sun and his people to leave the land of their ancestors, a village called "The White Apple." Located near the mouth of Second Creek, approximately 12 miles south of the fort, the village spread over an area of about 3 miles. The Great Sun made several attempts to reason with Chopart, all of which fell on deaf ears. To the great Natchez king, this left only one option.

Armed with knives and weapons, the Great Sun and his warriors approached the fort that November morning posing as neighbors wishing to procure a supply of ammunition for a hunting excursion by bartering with poultry and corn. Once inside, at the signal of their king, the warriors began a brutal attack in which most of the fort's inhabitants—men, women, and children—were killed. As the smoke began to rise from the fort, other parties entered and joined in the attack. While the massacre proceeded, the Great

Sun seated himself in the French warehouse and smoked his pipe as the warriors dropped the heads of their victims in a pile at his feet. Within three hours, almost every male inhabitant of the colony was dead. Only slaves, a few women and children, and those known to have useful skills such as carpenters and tailors were spared.[4]

Andre Penicaut, a ship's carpenter, had accompanied Pierre Le Moyne d'Iberville's second expedition to Mississippi in 1699 at the age of 19 aboard the vessel *Le Marin*. He established himself in Mobile and later explored the new continent, possibly as far north as Minnesota and west to Texas, although some scholars doubt Penicaut's claims of the extent of his travels. The fact that Penicaut spent much time at Natchez from 1704 until 1721, however, is not in dispute. Here, he became a student of the Natchez people. Going blind in 1721, he returned to his native country, where he wrote a manuscript, "Annals of Louisiana," which made its way into the king's library in Paris. Penicaut's writings provide a valuable record of early life in the region, including an accurate description of the deteriorating relationship between the Natchez people and the French colonists. In addition, he would call upon his time with the Natchez to describe a nut previously unknown throughout most of Europe.

Penicaut reports that upon his arrival at the village of Natchez in 1704, the Natchez people had three kinds of walnut trees, one of which bore nuts as large as a man's fist, from which the Indians made bread for their soup. A second tree Penicaut described as producing nuts not much larger than a thumb. These, which the Indians called *pacanes*, he described as the tastiest of all.[5] Ironically, it was the subsequent misfortune of Penicaut's blindness that may have spared his life so he could return to France several years before the massacre at Natchez, allowing him to enlighten the rest of the world about the pecan's existence.

The French also found the pecan farther up the Mississippi River to the north. Father Gabriel Marest's journal entry of November 9, 1712, describes "*les pacanes*," the fruit of a nut tree that "have a better flavor than our nuts in France," in the Kaskaskia region along the Mississippi River near the mouth of the Kaskaskia River in Illinois. Another French missionary, Xavier

Charlevoix, described the pecan in his diary entry of October 21, 1721: "Among the fruits that are peculiar to this country the most remarkable are the pecan. The pecan is a nut having the length and form of a large acorn. There are those with a very thin shell. . . . All have a fine and delicate taste." While the name is often attributed to the Algonquin language, it is not known whether "pecan" truly is derived from the Algonquin, or from the Natchez, or simply from a term commonly used among many Native American groups. We do know, however, that from the French spelling of the Native American word *pacane*, we have derived the word "pecan."[6]

While the spelling is pretty much universal today, pronunciations of "pecan" can be quite varied. Some folks say "pee-can," some say "puh-khan," and some even say "pee-khan." I once heard a farmer explain the difference between pee-cans and puh-khans: "When the crop brings two dollars per pound, they are puh-khans and when they bring fifty cents per pound, they are pee-cans." Another colorful Southern commentary on the pronunciation of pecan is "a pee-can is a can you keep under your bed for emergencies at night; a puh-khan is a nut that you eat."

All plants known to science have been given latinized scientific names, in addition to their common names. Unlike the common name, which can vary from country to country and even between regions, the scientific name is internationally accepted and does not change, at least once it is settled upon. Scientists have historically had a difficult time deciding what to call the pecan. The first recorded scientific name of the pecan was *Juglans illinea*, given by Richard Weston, an English botanist, in 1775. Weston, although employed as a thread-hosier in Leicester, England, fancied himself a "country gentleman." He was regarded as having a very wide knowledge of plants and plant literature and served as secretary of Leicester's agricultural society. Weston published his name for the pecan in *English Flora*, a list of epithets he developed along with their English common names, arranged by family. He referred to *Juglans illinea* as the Illinois walnut tree. However, plant taxonomists refer to Weston's name for the pecan as a "nomen nudum," or naked name. This term is used to describe a name or phrase that looks like a scientific name and may have been intended to be a scientific name, but fails to be recognized as such because it was not published with an adequate description of the organism and is thus "bare" or "naked."

The next official moniker given to the pecan was *Juglans pecan*, a name given by Humphrey Marshall, a Pennsylvania Quaker botanist and younger cousin to the famed botanist John Bartram, in 1785.[7] Published only two years after the formalization of American independence, Marshall's *Arbustrum Americanum* is the first botanical treatise on American plants written and produced by an American.

At the age of 12, Marshall was apprenticed to a stone mason, but his avid interest in nature led him to other pursuits. By the time he had reached his 20s, Marshall's botanical skills were recognized by the scientific community and he became a much-sought-after supplier of native plant and animal specimens. Eventually, Marshall corresponded with some of the leading botanists in England and America, including John Fothergill, Sir Joseph Banks, Benjamin Franklin, Timothy Pickering, Hector St. Jean de Crèvecoeur, and Johannes Fredericus Gronovius. By 1764 Marshall had constructed a rare plant conservatory on his farm in Chester County, Pennsylvania. Within 10 years Marshall's farm had developed into an export-oriented botanical garden stocked with local flora and exotic plants from throughout the United States and Europe.

Marshall was a prolific publisher of scientific papers. He produced works on the natural history of tortoises, sunspots, and agriculture. But it is for the *Arbustrum Americanum* that he is best remembered. This book is dedicated to Benjamin Franklin and the other members of the American Philosophical Society. Although not specifically devoted to the southeastern United States, Marshall's work is considered a valuable contribution to the botanical study of the region.

The problem with Marshall's name for pecan was that his description was inadequate to distinguish it from another species, *Carya cordiformis*, the bitternut hickory. So, in 1787, German botanist Julius von Wangenheim gave the pecan the name *Juglans illinoinensis*. Wangenheim's drawings show that he used poorly filled kernels in his description of the fruit, which are not uncommon under conditions of drought or environmental stress on the tree.

Karl Koch was another German botanist, best known for his explorations in the Caucasus region of Eurasia. Sadly, most of the botanical collections that made up his life's work have been lost to history. Koch transferred the species from the *Juglans* to the *Carya* genus, apparently changing the species name from *illinoinensis* to *illinoensis*. Why this was done, no one is quite sure. Most believe it was a simple spelling mistake. Koch made no formal mention of his alteration of the name and consistently used *illinoensis* throughout his description of the tree. He made several other errors in reference to Wangenheim's work, including several dates and figures.

This problem led to a great deal of confusion among scientists as to which scientific name to use when referring to the pecan. Some camps spelled the Latin name for pecan *Carya illinoensis*, while others used the spelling *Carya illinoinensis*. In 1964, the spelling *C. illinoinensis* was rejected without much explanation. After much debate, *C. illinoensis* was submitted to a committee (the Standing Committee for the Stabilization of Nomenclature) that decides such things as the correct spelling of a scientific name.[8] The proposal was rejected, finally lending the pecan a definitive name in the scientific literature, *Carya illinoinensis*.[9]

Pecans and Humans

Regardless of what they are called, pecans have a unique history among nut crops, in that their appearance and range were probably largely shaped by a close association with humans. The family to which pecans belong—Juglandaceae—is an old family that arose in the Cretaceous period, about 135 million years ago.[10] The world at that time would be nearly unrecognizable today. There were two great supercontinents: Laurasia—a fusion of North America and Eurasia to the north—and Gondwana, made up of Antarctica, South America, Africa, Madagascar, Australia, New Zealand, Arabia, and India, to the south. The two great continents were separated by an equatorial seaway.

There was little or no ice at the poles and any land along the equator was arid, as opposed to the humid tropical rain forests of today. The Arctic was covered with lush, green forests rather than ice, and dinosaurs roamed the land. The seas reached their highest levels in history, flooding much of the land that is today dry.

One of the most significant aspects of the Cretaceous environment was that global temperatures were approximately five degrees warmer than they are today. It was under these conditions that flowering plants arose, among which, of course, was the Juglandaceae or walnut family. The "hickory" family branch to which the pecan belongs—the Carinae—is believed to have evolved from the primitive members of the walnut family about 70 million years ago. The early hickories eventually came to be distributed across North America and Eurasia. The oldest specimens of "hickory-like" fruit, dating back about 34 million years, were recovered from places as far apart as Colorado, Germany, and later China, as a result of the cleaving of the great land mass of Laurasia millions of years ago, which distributed the hickory family around the world. As glaciation reached Europe about two million years ago, the hickories disappeared from the continent but remained in North America and China. At the same time as their European extinction, hickory diversity in North America was greatly reduced and they vanished from the North American west.[11]

Biologists consider the place where a species has the widest array of forms to be its "center of diversity," or its ancestral home. For instance, there are hundreds of varieties of corn or maize in Mexico that are found nowhere else on earth. Thus, maize is recognized as originating in Mexico. Because the greatest diversity of species within the walnut family occurred in North America during the time of its radiation across the globe, it is believed that this continent is most likely the place in which the family originated. The earliest fossil "hickory-like" fruits lacked secondary septa or partitions in the walls of the nuts. As hickory species evolved, they developed secondary septa and thickened shells as a means of defending themselves against rodents. The one exception to this case was the pecan, possibly as a result of human intervention in the process of natural selection.

I grew up, like many in the South, assuming that pecan trees were native to the southeastern United States. Pecans now commonly grow outside their native range along fencerows and field edges throughout Alabama, Georgia, northern Florida, North Carolina, and South Carolina. They occur here now in such abundance and seem to fit so well as part of our natural landscape that it seems only right that pecans should have always been here. Indeed, the depths to which the pecan tree has become ingrained into the culture of this region are reflected in its art and literature. Harper Lee's classic tale *To Kill a Mockingbird* even has a passage devoted to the pecan: "from the Radley chicken yard tall pecan trees shook their fruit into the schoolyard, but the nuts lay untouched by the children: Radley pecans would kill you."[12] Still, although the southeastern United States is now covered with pecan trees from yards to orchards, the tree is not native to the area.

The "natural" distribution of pecan is a murky subject, made so by the affinity of humans for this nut through the ages. The native range of pecan is believed to be along the Mississippi River and its tributaries from as far north as Clinton, Iowa, southward to the Gulf Coast. The bottomlands along the tributaries of the Mississippi extending east and west are also lined with pecan southward to the Gulf of Mexico. The pecan has been reported up the Ohio River as far as Cincinnati and upriver on the Wabash to Terre Haute, Indiana. Reports also exist of the pecan being found up the Tennessee River as far east as Chattanooga, but this potential native range is looked upon with suspicion. Most of the large pecan trees occurring near Chattanooga show evidence of being planted by humans.[13]

The pecan is found in central Oklahoma along the Arkansas River, southward along the Red, Brazos, and Colorado Rivers, and along portions of the Rio Grande in Texas and into Mexico. The pecan reaches westward to the Edwards Plateau, an elevated region extending west into central Texas and formed from marine deposits of sandstone, limestone, shale, and dolomite 100 million years ago when the region was covered by ocean. Most of the streams of this area are bordered by pecan trees that lend shade as well as nuts for food. Pecan is also found in regenerating isolated stands as far south as Zaachila,

Oaxaca, Mexico. Isolated populations of pecan also occur in southwestern Ohio, northern Kentucky, and even central Alabama.[14]

The northern and western boundaries of the pecan are easily explained by climatic conditions. At the northern end of the range in southern Iowa, pecans are limited by the number of frost-free days (over 180) required for the completion of their life cycle. The trees will grow in areas receiving fewer frost-free days but usually do not produce fruit. In the west, soil moisture limitations on tree growth are encountered. The eastern boundary of the pecan presents a greater problem. Although soil moisture is adequate for pecans in the east, the soil's acidity increases as you move farther into that region. Adapted to the well-drained, loamy, more alkaline soils of the river bottoms along the Mississippi and into Texas, the pecan cannot compete with other species in the east that are more tolerant of acid soil conditions.

The mysterious isolated populations of Alabama, Texas, and Mexico are the primary source of speculation centered on the role that humans may have played in the introduction of the pecan to these areas so far removed from its known natural range. Early settlers and Indians traveled great distances carrying pecans in their pockets and pouches. Although at first glance, it seems likely that pecan was probably brought into these areas by humans, the evidence is not clear cut. Native stands of pecans are recorded along the Black Warrior River in Alabama from around the turn of the twentieth century.[15] Some believe that these trees originated from nuts planted by the Choctaw Indians. These stands of pecans are located near Moundville, an important site of the Mississippian culture of the southeastern United States, where archaeological excavations confirm that pecans were present as early as 1050 to 1250 AD.[16] Records from the late eighteenth century document the trading of pecans from the lower Mississippi River Valley, which also casts a shadow of doubt on the native range of the species in southern Mississippi and Alabama.

Other species, such as nutmeg hickory and shagbark hickory, are found in similar isolated or "disjunct" populations in Mexico, far from their areas of major occurrence in the southern United States. Because of their apparent lack of food value to humans, it is believed highly unlikely that humans introduced these two hickories to isolated regions. Therefore, many scientists believe it reasonable to assume likewise for the pecan.

One factor confounding the ability to determine the true historical range of pecan is the inability to distinguish pecan pollen from hickory pollen.[17] Our knowledge of the ancient distribution of plant populations is often based on pollen recovered from soil cores. As the depth at which the pollen is found in a core increases, so goes the age of the pollen. From pollen studies, we can say with relative confidence that hickories in general spread rapidly northward about 16,000 years ago, with the period of greatest northward advance occurring between 12,000 and 14,000 years ago. The current northern limit of the hickories is at about 45° north latitude. But how does the pecan fit into this scenario? Where is the pecan's center of diversity?

The short answer to this last question is, we don't know. L. J. Grauke, a horticulturist at the USDA's pecan station in Somerville, Texas, is getting close to an answer using molecular analysis of the plant's DNA. Scientists group organisms with similar DNA patterns into "haplogroups." Those belonging to the same haplogroup are considered to share a recent ancestor. Conversely, those individuals that do not share a recent ancestor belong to separate haplogroups. Grauke's analysis of the pecan's DNA pattern found that Mexican pecan populations have nine haplotypes, six of which are found only in Mexico.[18]

Scientists can examine DNA patterns in several ways. One way is by examining the DNA of plastids in the plant. Plastids are organelles in plant cells that are ultimately responsible for the production of food via photosynthesis. Chloroplasts are an example of one type of plastid. In the hickory family, plastid DNA is inherited only from the female parent.

Another method of studying DNA is to analyze nuclear DNA. This is what most of us think of when we hear the term "DNA." Nuclear DNA is found in the nucleus of every cell of an organism. It encodes more of an organism's genome than does plastid DNA in that nuclear DNA results from the "mixing" or recombining of genes from both the male and female parents, or in the case of a pecan tree, from the seed parent and the pollen parent.

Unlike nuclear DNA, with genes rearranged in the process of recombination, plastid DNA usually carries no change from parent to offspring. Because of this consistency, plastid DNA can be a powerful tool for tracking maternal ancestry. As a result, it has been used to track the genetic lines of many species back hundreds of generations.

According to Grauke, plastid diversity is greatest in the Mexican populations, but nuclear diversity is greatest north of the Mexican border, in what we know are the "youngest" populations. The high nuclear diversity found in the United States may simply be the result of a lot of hybridization, nature's way of rapidly increasing diversity. The plastid diversity found in Mexico's native pecans might be based on these southern populations being refugia where trees, or rather their lineage, have survived from the Neogene period. In this scenario, the retreat of glaciers and land use may have isolated these populations, formerly connected to each other in an ancient forest. The isolation would have allowed them to become well adapted to their locations over time. Alternatively, the pecan diversity found in Mexico could have resulted from dispersal from northern populations. However, so far, Grauke has found no populations in Texas that could have been the source of the southernmost Mexican populations in Oaxaca and Ixmiquilpan. To date, the evidence points to the Mexican populations being the oldest. Still, Grauke suggests, the enormous genetic pool of pecans in native stands throughout their range and the lack of researchers to decipher this puzzle create large gaps in the information.[19]

Pecan remains have been discovered in a number of prehistoric sites throughout North America, including Illinois, Texas, and Louisiana. These recoveries tell the story of the earliest known presence of pecan in North America and help demonstrate the value of the nut to Native American cultures. From these archaeological sites, we know that humans were using pecans as early as 6750 BC.

Grauke's molecular analysis suggests that pecans and other hickories found in the northern end of their range originated not directly from Mexico, but from populations in the Southeast after the retreat of the Ice Age glaciers.[20] One of the oldest-recorded pecan nuts from the northern Mississippi

River Valley was found in a geologic core from the floodplain of the Missis-sippi River near Muscatine, Iowa, in the late 1980s. The fossilized, intact pecan was found buried almost 10 feet below the surface and was estimated to have been deposited there over 7,000 years ago.[21] Pollen records from the same gen-eral area suggest that hickory pollen was abundant in the region from 10,000 to 7,000 years ago, accounting for as much as 34 percent of the total tree pollen there at that time. As one would expect, the pollen data indicate that more hickory species were growing along the river valley than in the uplands. The only two hickories common to these floodplains today are shellbark hickory and pecan. So, it is assumed that one or both of them are responsible for the hickory pollen found in the region's soil cores. But how did the trees responsi-ble for this pollen expand their range so far, so fast?

The most accepted answer to this question is that late Paleo-Indian and early Archaic people facilitated the spread of pecan up the Mississippi Valley about 8,000 to 10,000 years ago. Early Native Americans in the lower Missis-sippi and Illinois River Valleys used a variety of nuts, including other hickory species, acorns from several oaks, walnuts, butternuts, and hazelnuts, but the pecan was the most abundant nut at the time.[22]

The Mississippi and its tributaries were the "highways" of canoe travel for the Native Americans. As they traveled, the Indians are believed to have planted pecan nuts near their campsites. Some even suggest that as they did so, they may have planted the largest and thinnest-shelled pecans they collected, extending the range of the nut and helping to ensure the quality of pecans for future generations.[23]

The most significant spread of pecan in the Mississippi River Valley is likely related to the Native American Dalton Culture. Archaeologically speak-ing, the Dalton soil horizon is dated to between 10,500 and 9,900 years ago. At this time, hardwood forests were replacing the boreal forests of the Upper Midwest. The abundance of hickory pollen found in soil cores from the region at about the same time suggests that perhaps pecan and shellbark hickory were brought into the Upper Mississippi Valley, at least in part, by the Dalton peo-ple as they inadvertently dispersed the nuts while using them for food.[24]

Near the small town of Modoc in Randolph County, Illinois, there is a rock cliff undercut by floods that filled the Mississippi River Valley during the Ice Age. Native Americans took advantage of the rock shelter formed by the undercut cliffs as many as 9,000 years ago. The nomadic people who used this site lived by hunting animals and gathering plants for food and fiber. They regularly occupied the Modoc site off and on for more than 6,000 years. It is in the earliest period of habitation that most of the pecans are found at the Modoc site, accounting for as much as 5 percent by weight of the total nut usage at that time. Pecans are less commonly found in later periods, indicating a shift toward a heavier reliance on other hickories about 7,000 years ago after the development of processing practices, such as boiling, that allowed the hickory nuts to be more easily separated from their harder shells.[25]

The abundant stands of pecan trees along the creeks and rivers of southern Texas were central to the lives of Native Americans in the region and, as recorded by Cabeza de Vaca, were a seasonal stop in their annual migrations.[26] Oddly enough, although pecan trees blanket much of southern and eastern Texas, pecans are rarely unearthed in archaeological excavations in these areas now so dense with native pecans. This lack of pecan material is explained by several factors. First of all, where pecan stands are most common along the lower reaches of Texas's major rivers, very few archaeological digs have been conducted. Additionally, only plant remains that have been charred by fire, usually in some sort of processing, are capable of survival in open archaeological sites more than a few hundred years old. Pecans, as such, are rarely charred by fire, making them difficult to find in dig sites. Lastly, because pecan shells are so thin, they likely deteriorate relatively quickly compared to most other plant materials.

One of the most valuable regions for deciphering the history of the human relationship with pecan lies along the Edwards Plateau of west-central Texas. This region, often referred to as the Texas Hill Country, is covered by shallow soils located over limestone beds east of the Pecos River and west of the Colorado River. This large area is composed of various habitats, including oak-pine savannas, grasslands, canyons, woodlands, and forests. Roughly outlined by the cities of San Angelo, Austin, San Antonio, and Del Rio, the Edwards Plateau is an arid region. The only significant supply of water here is found along the region's streams. For this reason, travel across the plateau remained difficult until modern means of tapping underground water supplies were developed. As late as 1950, there was no railway line crossing the entire width of the Edwards Plateau. Because of the harsh conditions, ancient civilizations in this region tended to congregate around streams.[27]

Native pecan bottomland communities exist here only as fragments of the river systems. Although the pecan tree is a minor species throughout many Southern floodplain forests, nowhere else on earth does the pecan dominate as it does along the streams of the Edwards Plateau, where the tree exists in nearly pure stands.

It is mostly here, in the arid canyonland regions along the margins of the Edwards Plateau, that the few archaeological sites in which pecans have been recovered are found. Baker Cave is a large rock shelter located in a side canyon of the Devils River near the southwestern edge of the Edwards Plateau in the Lower Pecos region near Del Rio, Texas. This unforgiving region of rough, desolate terrain is one of the most significant archaeological regions in North America. Because of the unique ecological and geographical features found here, the canyonlands of the Lower Pecos harbor a rich cultural legacy—including 5,500-year-old cave paintings, grass-lined beds, painted deer bones, and even perfectly preserved food remains—rarely found elsewhere in North America. The rivers and their tributaries have cut deep canyons into the land over the centuries. Within these canyons, natural overhangs have formed along the soft limestone cliffs. The shelters and the reliable water from the rivers, numerous springs, and temporary water holes called *tinajas* made the area suitable for habitation by the ancient people that lived here for over 13,000 years.

Thick layers of pecans have been found in Baker Cave, the earliest of which dates back 5,000 years, possibly representing the western limits of pecan in the region.[28] Pecans have turned up in several other sites across central Texas, including the Kyle site, near the Brazos River in Hill County, Texas, where pecans were found in 650-year-old archaeological deposits. The Varga site, a 6,000-year-old campsite, is split by a paved road in Edwards County, Texas. The site lies just below the top of the Edwards Plateau along Hackberry Creek, a tributary of the Nueces River. Originally, surveyors found burned rock and chipped-stone artifacts scattered along Hackberry Creek. In 2001 and 2003, road reconstruction led to more extensive investigations after heavy rains and flooding damaged the road. Here pecan nut fragments estimated to be 350 years old have been found. Although it seems likely that pecans would have been used as food here throughout the thousands of years of sporadic use of the campsite, evidence is lacking because of the rapid deterioration of organic materials in this environment.[29]

So, while we have clues to the extent of the area in which pecans naturally grew in North America, the current technological limitations of science and the historical meddling of humans make it difficult to pinpoint the pecan's true native range without human interference. Perhaps we should consider this all-important relationship between humans and the pecan as simply a part of the life history of this tree. Without question, humans can be considered the primary dispersal agent of the pecan. Settlers cut down many native groves in the alluvial floodplains as they cleared land for farming. They also felled many trees for lumber to be used as building material during the early days of settlement. They even cut them down or lopped their branches off when they were hungry and grew tired of waiting for the nuts to fall.[30] The life history of the pecan is woven too tightly with the history of humans in North America to disqualify our interrelationship from any discussion of the pecan's origin.

Pecans were used not only as food by the tribes of the Mississippi River Valley and into Texas. The fermented, milky powcohiccora drink was consumed during religious ceremonies and tribal celebrations to enhance a warrior's bravery in battle. Most likely, powcohiccora was made from multiple species of hickory, of which the pecan was one.

Food and drink were not the only uses the Indians found for the pecan. The Comanche were known to grind pecan leaves and use them as a poultice for treating ringworm. The Kiowa administered pecan bark to those suffering from tuberculosis. Because pecans were so important to the life and culture of certain Native American tribes, they tended to revere the pecan tree. The traditional role and value of the pecan to Native Americans is well illustrated by an ancient tale of the Caddo Indians of eastern Texas, Oklahoma, and Louisiana. "The Old Woman Who Kept All the Pecans" was documented by George Dorsey in his book *Traditions of the Caddo*:

> There lived an old woman who was mother to all the pecan trees. She owned all of the trees and gathered all the nuts herself. When people went to her lodge she would give them a few pecans to eat, but would never allow them to take any away. The people were very fond of pecans and they wanted some for their own use, but the old woman would not let them have any. One time the people were very hungry and the old woman had everything in her lodge filled with pecans, but she would give them only a few when they went to see her and she made them eat them before going away. This made the people angry and they decided that something must be done. There was in the village an old man who had four little sons who were very troublesome and meddlesome—they were the field Rats. The people thought that these four little boys would be the right ones to go over to the old woman's house some night to try to steal some of the nuts. They chose the four boys both because they were small and quiet and sly and because they were such a nuisance around the village that they would be no great loss to the people if the old woman killed them. The Rats were willing to go because they were always glad to be meddling. They chose one to slip over and make sure that the old woman was asleep. He went to her lodge and peeped in through a small crack and saw that she was still at work. He waited until she finished her work and went to bed; then when he heard her snore he ran back home to tell his brothers to come. When he went inside his father's lodge, he saw a stranger sitting there. The stranger was Coyote. He had come to tell the Rats not to trouble about stealing pecans from the old woman, for he was going over the next day and kill her. Coyote was afraid to trust the Rats. He wanted to go himself, so he could get most of the pecans. The next morning he

went over to see the old woman and acted very friendly. The old woman gave him some pecans and he sat down and ate them all up. Then he asked her for some more, and as she turned around to get them he pulled out his stone knife and struck her on the head. She died, and ever since then the pecan trees have grown everywhere and belonged to all of the people.[31]

What made the pecan such an important food for the Indians? Granted, in such a harsh environment, food was likely hard to come by. The Indians had no pack animals with which to transport their food stores. Many of the root crops they used were low in energy compared with the pecan and required extensive processing for storage in order to remove the excess water and concentrate the carbohydrates. The pecan, by contrast, came readily packaged in a handy shell and could be easily accumulated, concentrated as they were in known locations among the groves of the river bottoms. In addition, pecans were full of energy, providing 19 vitamins and minerals, and an astounding 690 kilocalories per 100 gram of nutmeat. Because of its low water content, the nut required no processing before eating except the simple removal of the meat from the thin-walled shell.[32]

Cabeza de Vaca's own account tells us that the Mariame tribe along the Guadalupe River ate pecans for one to two months out of the year, and their groves attracted other groups from as far as 75 miles away to take part in the harvest. The Iguace Indians along the Texas coast rarely ate deer or fish. Instead, they preferred various roots and the fruit of the prickly pear cactus. In winter they traveled northwest to the Colorado River, where they ate pecans and little else throughout the season.

Juan Sabeata, a Jumano Indian living in western Texas from 1683 to 1692, gave an account of the willingness of the bands to travel great distances for the pecan harvest and of this seemingly jovial time during the cycle of the Indians' year: "The river which they call Las Nueces is a three day's journey from that place; that there are nuts in such abundance on this river that they constitute the maintenance of many nations who enjoy friendship and barter and exchange with his." The Mescalero Apache followed the bison herds southward from the Sacramento Mountains and Sierra Blanca of New Mexico to the pecan groves along the upper Colorado and Concho Rivers. They maintained camps throughout the winter months along the San Saba, Pedernales, and Llano Rivers in central Texas, where they gathered the abundant pecans and hunted buffalo.[33]

Indians no doubt learned quickly of the tendency of the pecan tree to bear fruit in biennial cycles, or in general about every other year, a phenomenon often referred to today as alternate or biennial bearing. Many accounts report the Indians' knowledge of the tree's alternate bearing characteristics. By necessity keen observers of their environment, Indian peoples had knowledge of the land that would also have helped them keep track of which trees were good producers.

Compared to hickory nuts and other nuts common to eastern US forests, pecans were much easier to gather. Wild pecans in Texas, Oklahoma, and Louisiana have been shown to grow in equal or greater densities per acre than hickory and have a much greater yield per tree. In 1709, Espinosa wrote of the Payaya Indians inhabiting the area from the Medina to the San Marcos Rivers: "The nuts are so abundant that throughout the land the natives gather them for food the greater part of the year. For this purpose, they make holes in the ground where they bury them in large quantities. . . . The Indians are very skillful in shelling them, taking the kernels out whole. Sometimes they thread them on long strings, but ordinarily they keep a supply in small sacks made of leather for the purpose" It has been suggested that the underground pits and leather pouches were a means of storage used by the Payaya Indians as a hedge against predictable light crops the following year. The average collection rate for pecans would have been about 13.5 pounds per person per hour. It is known that certain trees within some native pecan groves yield appreciably larger nuts, so that collecting beneath them yields larger returns per unit of collection time. Using hammer stones and rock anvils to crack native pecans, an average of about 66 grams of nut meat per person per hour would have been expected.[34]

The Native Americans valued the pecan not only for consumption but for trade as well. In the mid-seventeenth century, an expedition led by Hernan Martin and Diego del Castillo made its way into central Texas in order to gather freshwater pearls on the Concho River. While the pearls proved to be of little value, the expedition members established a valuable trade in buffalo skins and pecan nuts with the Jumano Apache, who were so pleased with the trading themselves that they transported the nuts from the river valleys of central Texas to the Rio Grande Valley. Trade between the Jumano and the Spanish was interrupted from 1680 to 1683 when the Pueblos revolted. Following the revolt, a group of Jumano approached the Spanish governor to request a resumption of the trade of pecan nuts. After 1617, New Mexico's Spanish missions were outfitted by supply expeditions from Mexico every three years. It is speculated that perhaps pecans made their way back to Mexico from the nuts traded by the Jumano Apache on these supply trips.[35]

As the Cherokee were forced on their long march along the Trail of Tears in the 1830s, they resettled near the Creek tribes along the Arkansas River. Here, they cleared land to plant crops, but recognizing the pecan, they spared groves of the trees, from which they gathered nuts as food and trade items. Colonel James Logan, a Creek Indian agent, noted, "The sale of pecan nuts, the trees bearing which abound in the rich bottom of the watercourses, is of considerable importance to this class. It is estimated that the quantity sold to the different traders during the last fall and winter amounted to between 9 and 10,000 bushels, the price for which was from 50 cents to $1 per bushel, and was generally bartered for necessary articles of Clothing, Sugar, Coffee, & Salt, &c, besides a large quantity was no doubt used for food."[36]

Indians were not the only ones to exploit the value of pecans as a trade item in the new frontier of the nineteenth century. In the 1830s along the Brazos River, James Perry harvested and sold wild pecans from the bottomlands of his Peach Point Plantation just west of Houston. Ben McCulloch, born in Tennessee and reared near Dyersburg, lived not far from his neighbor David Crockett. As Crockett prepared to go to Texas, McCulloch planned to meet him in Nacogdoches on Christmas Day 1835. As fate would have it, however, McCulloch contracted measles, which slowed him down and ultimately

saved him from perishing alongside Crockett at the Alamo. After serving under General Sam Houston in the battle of San Jacinto, McCulloch eventually
went on to further distinguish himself as a Texas Ranger, US Marshal, and
brigadier general in the Confederate Army. During a rare lull in the action in
1837, McCulloch settled along the San Marcos and Guadalupe River Valleys.
Here, along with his brother and two other men, McCulloch built two flat-
boats to float down the Guadalupe River of Texas. They loaded the boats with
pecans and set off downstream to Saluria Island at Pass Cabello, where they
sold and traded nuts. This was McCulloch's only foray into the pecan trade.
Perhaps it was a bit too tame for a man who craved the type of action McCulloch saw.[37] However, the trade of pecans was likely a foundation of the early
Texas pioneer economy.

DISCOVERY

Pecans have been connected with three of the most important Spanish explor-
ers of North America; Cabeza de Vaca, Coronado, and de Soto. Following
Cabeza de Vaca's journey across North America, the Spanish continued to explore this new land and on their many journeys continued to note the pecan.
In 1539, Francisco Vásquez de Coronado y Luján, governor of New Galicia, a
province of New Spain located in the contemporary Mexican states of Jalisco,
Sinaloa, and Nayarit, dispatched two men on an expedition north toward New
Mexico. Only one of these men, Marcos de Niza, returned with an incredible
story about a golden city of great wealth called Cíbola. According to de Niza,
his comrade Estevanico had been killed by the Zuni citizens of Cíbola. The
surviving de Niza claimed not to have entered the city of Cíbola. Instead, he
viewed it from a distance as it stood on a high hill. From his vantage point
the city appeared to be made of gold. Off to the west, he could see the Pacific
Ocean. As any conquistador worth his mettle would, Coronado became ob-
sessed with the city of gold and planned an expedition to find it. On his quest
for the mythical city, Coronado trekked through present-day Arizona, New
Mexico, Texas, Oklahoma, and Kansas from 1540 to 1542. After finding the

Zuni Indians, Coronado was vastly disappointed that no city of gold seemed to exist. A scouting branch of the expedition did, however, discover another treasure, the Grand Canyon. On his return trip, Coronado searched for a rich land called Quivira and met the Teyas Indians as he descended the tabletop of the Llano Estacado, most likely in Blanco Canyon along the White River in present-day Crosby County, Texas.

The Teyas inhabited the canyons of the eastern panhandle of Texas and the adjoining region of Oklahoma, which housed flowing streams with trees along their banks. The Teyas were primarily buffalo hunters whom the expedition's chroniclers noted did not "sow corn, nor eat bread, but instead raw meat." Coronado also noted the Teyas's use of mulberries, roses, grapes, plums, and nuts, the latter of which were most likely pecans. Coronado had one intriguing encounter with an elderly member of the Teyas tribe who had a beard, a rare occurrence among the native people. The man told Coronado that he had met four Spaniards far to the south over a decade earlier. These could only have been Cabeza de Vaca and his fellow survivors.[38]

Hernando de Soto traversed the southeastern United States between 1539 and 1542, passing through what is today the heart of southeastern pecan-production territory from Florida to Louisiana until he died of fever along the banks of the Mississippi River in present-day Arkansas. The chronicle of his expedition described "walnuts in great number of a new type bearing soft shelled nuts in form like bullets, which the Indians laid up in great stores in their houses." He also described the trees themselves as being different from those of Spain in that they had smaller leaves. There seems to be no doubt that what de Soto was describing were pecans, but there is some discrepancy as to the location of his observation. He describes it as lying within the country of the chief called Casqui. This was originally interpreted to be an area known to the French as Kaskaskia, the area earlier described by Father Marest and Charlevoix. Although de Soto is not thought to have actually traveled this far north, he again described pecans as occurring along the Washita River in present-day southern Arkansas, as well as in a location just west of the mouth of the Yazoo River in northern Louisiana.[39]

The Kaskaskia region became well known for the presence of the pecan. Edmund Flagg, in the mid-1830s, also reported herds of wild horses, plums, persimmons, cherries, "the delicate pecan," hickory, and hazelnut near Kaskaskia. He describes "crossing Cahokia Creek . . . and threading a grove of pecan, with its long trailing boughs and delicate leaves." Native to Maine, Flagg was a journalist during this period of his life. His travel articles for the *Louisville Journal* are filled with dramatic descriptions of the westward-expanding American continent. As he looked back over his shoulder along the banks of the Mississippi, leaving Saint Louis, Flagg described the pecan tree's native land as "the closest thing to a cosmorama . . . the American Bottom . . . a tract of country which, for fertility and depth of soil, is perhaps unsurpassed in the world."[40]

Over the 100 years following de Soto, explorers of the Mississippi River region and Texas constantly referred to walnuts or *nogales*, which many believe imply pecans. From 1683 to 1684, Pedro de Mendoza explored parts of present-day Texas along the region of the junction of the Concho and Nueces Rivers and farther east to the Colorado. Here he describes groves occurring along the bottomlands of the watercourses and great groves of very tall pecan and live oak trees in another location.[41] Marching from Mexico into Texas in 1689, Ponce de León also described seeing pecans, which he referred to as walnuts (the only thing the Spanish had to compare them with) or *nogales*, along the Nueces and Medina Rivers and Atascosa Creek.[42] For all their exploration from the fifteenth through the eighteenth centuries, neither the Spanish nor the French appear to have sent the pecan back to their homelands, or if they did, it failed to make a lasting impression. Although many descriptions of the pecan by these explorers exist, there are no verifiable records of its introduction into Europe during that time. It was the mid-eighteenth century before the pecan would penetrate the English colonies and Europe would come to know the pecan.

John Bartram, eighteenth-century botanist, naturalist, explorer, and horticulturist, was born a Quaker on a farmstead in Pennsylvania. Having no formal education, Bartram considered himself "a plain farmer," yet he is now considered the "father of American botany." Bartram had a voraciously curious mind and read widely, having a keen interest in medicine and medicinal plants. His botanical career began with the planting of a small area of his farm with a few interesting plants. After developing relationships with European botanists interested in North American plants, Bartram turned his hobby into an occupation. He traveled the East Coast from Canada to Florida documenting the plants, animals, climates, geology, and rivers of the New World. Many of the plants he found were transported to collectors in Europe in exchange for books and supplies. Bartram's eight-acre botanical garden on the banks of the Schuylkill River near Philadelphia is considered the first true botanical collection in North America.

On August 14, 1761, Bartram wrote his friend and fellow plant lover, English merchant Peter Collinson, "I have not yet been at the Ohio, but have many specimens from there. But in about two weeks I hope to set out to search for myself, if the barbarous Indians don't hinder me (and if I die a martyr to botany, God's will be done)." It is believed that on this trip, Bartram made his first collection of pecans. These pecans may very well have been the first to reach the Atlantic Seaboard, although according to one of the earliest US Department of Agriculture texts to discuss the pecan, *Nut Culture in the United States*, the pecan was unknown on the Atlantic Seaboard until 1762, when fur traders from the Mississippi Valley brought the nuts to New York. However, strong evidence exists for Bartram's pecans being the first to make it to the "civilized" East Coast.

Bartram mailed a box of plants and seeds on December 12, 1761, to Collinson, who often distributed the novelties from Bartram's boxes to a long list of European clients. It is likely that this box contained the first pecans to reach the shores of England. Collinson's return letter to Bartram on April 1, 1762, indicates the puzzlement of the European botanical community over these "stony seeds": "I really believe my honest John is a great wag, and has sent me seven hard, stony seeds. Something shaped like an acorn, to puzzle us; for there is no name to them. I have a vast collection of seeds, but none like them. I do laugh at Gordon, for he guesses them to be a species of Hickory." Bartram's reply states the following: "The hard nuts I sent were given me at

Pittsburgh by Colonel Bouquet. He called them Hickory nuts. He had them from the country of the Illinois. Their kernel was very sweet. I am afraid they won't sprout, as being a year old."[43]

John Bartram's son William, noted early naturalist and explorer of the southeastern United States from 1773 to 1778, regarded the pecan as "one of the most useful trees of the country." In November 1777 he made note of two large trees of the *Juglans* pecan growing in a Mobile garden. In addition, the younger Bartram often referred to "*Juglans*" of various sorts along his trip from Mobile to the Mississippi River and Baton Rouge.[44]

Early Cultivation

Although American Indians often used pecans as a food and trade item, we have no hard evidence that they actively cultivated the tree. The oldest-known cultivated planting of pecans was made in the late 1600s or early 1700s near the town of Bustamente in northern Mexico. These trees were likely either planted as seeds in the spot on which they grew or were grown from seeds and transplanted. Documentation of this early planting of pecans is dated to as far back as 1711, approximately 70 years before the earliest recorded planting in the United States.[45] Lying within a canyon of sheer rock faces, Bustamente was settled in 1686 by 30 families of Tlaxcalan Indians who had applied for land and water rights in the area. In the late seventeenth and early eighteenth centuries, Bustamente underwent an economic boom thanks to the waters flowing from the springs at the west end of the canyon. The settlers diverted the springs into three acequias, small earthen irrigation ditches that fed the garden community, allowing them to plant orchards of avocados and pecans, the harvest of which they sold to the mining towns of Villaldama and Lampazos. Water was the key to this earliest recorded planting of pecans. J. H. Burkett, a nut specialist with the Texas Department of Agriculture who visited Bustamente in the 1920s, found pecan trees dead or dying in the plots of small farmers after they sold their water rights and could no longer irrigate their trees. The 75-year-old trees, planted in very shallow soil underlain by solid rock, could not bear the dry desert conditions and perished of thirst when their roots went searching deep, only to find rock below. Today, the core of each urban block in Bustamente is a garden of towering pecan and avocado trees, providing shade to residents and visitors, thanks to the acequias still running through the plots and beneath the streets of the town.

The Spanish missions established along the Rio Grande by the Franciscans in the seventeenth century often established their own farms on which they grew corn, beans, cotton, and chili peppers. Some farms maintained orchard crops such as grapes, peaches, apples, pears, figs, olives, almonds, apricots, dates, and pomegranates. It would seem that the Spanish, well experienced in the production of orchard crops, would also plant pecan nuts throughout the valley. However, few records of pecan trees exist in the region. One large old tree was reported at Mesilla, just north of Las Cruces, which is now the center of southwestern pecan production. It was estimated to be at least 300 years old in the late 1960s and may represent a very early planting outside the tree's natural range.[46]

Although there are no known records to confirm its planting date, one of the most intriguing pecan plantings may be one reported to have occurred in Spain in the late eighteenth century. Noted botanist and explorer David Fairchild writes of finding several old pecan trees at Retiro Torre Molines in southern Spain. He describes the trees as being of large size, estimated to be at least 100 years old, and bearing small, hard-shelled nuts. Most likely, these trees would have grown from nuts that had been shipped from Mexico or Florida during the years of Spanish exploration; however, no evidence of such a shipment, aside from the trees themselves, exists to verify this theory.[47]

Prince Nursery on Long Island in Flushing, New York, was the first commercial tree nursery in the United States. Founded by Robert Prince in 1737, the nursery operated for 130 years, developing a plant import and export business with Europe. Initially called the "Old American Nursery," Prince's farm became a training ground for many early nurserymen and was the largest supplier of fruit trees and grapes in the New World, producing most of the grafted apple, pear, and cherry trees that could be found in early northeastern orchards. The British government deemed Prince Nursery so valuable that British general Lord Howe ordered its protection during the Revolutionary War. The nursery's first advertisement was dated September 21, 1767, and its first catalog was published as a broadside in 1771, featuring a wide selection of fruit trees, including 33 different plums, 42 pears, 24 apples, and 12 nectarines. By the late 1820s, Prince Nursery offered more than 100 species of Australian plants, and more than 600 types of roses. Following the expedition of Lewis and Clark from Saint Louis to the Pacific Ocean, many of the shrub and flower specimens the explorers collected were sent to Prince Nursery for propagation and distribution.[48]

The nursery's second proprietor, William Prince, is credited with developing his father's nursery business into the best plant nursery of its time in America. In 1772 William planted 30 nuts, the first recorded pecan planting in the United States. No one knows the source of those nuts, but 10 plants survived and germinated from their planting. Eight of these trees were shipped overseas and sold in England at a price of 10 guineas each, which would equate to nearly $1,400 US dollars today.[49]

One of the most notable early plantings of pecan in the United States occurred in 1775, when George Washington planted pecans at Mount Vernon. Washington is well known to have been an enthusiast of plants and agriculture. In fact, Washington considered himself to be first and foremost a farmer. Through the years he became one of the nation's leading agriculturists. Upon his retirement from public service following his second term as president in 1797, Washington wrote, "I am once more seated under my own vine and fig tree, and hope to spend the remainder of my days . . . in peaceful retirement, making political pursuits yield to the more rational amusement of cultivating the Earth."[50]

The grounds of Mount Vernon were full of plantings—field crops, flower gardens, vegetable gardens, experimental gardens, and fruit orchards. The primary crop at Mount Vernon was wheat, although Washington experimented with over 60 field crops and a variety of fruits. The first president also played a major role in introducing the mule to the United States as a beast of burden to aid in farming. He correctly believed the mule to be stronger and easier to care for than horses. Following the Revolutionary War, Washington received a donkey named Royal Gift as a present from the king of Spain and another named Knight of Malta from the Marquis de Lafayette. By breeding Royal Gift and Knight of Malta with his horses for work on Mount Vernon and distributing the offspring throughout the country, the first president helped increase the nation's stock of mules and establish them as reliable work animals. Washington also set about improving his land through crop rotation, fertilizers, plowing practices, and other new farming methods.

Washington spent much of his life experimenting with promising plants and crops, on which he kept notes in great detail. His diary entry from March 11, 1775, describes his first pecan planting in the experimental plots he called his "Botanic Garden": "Row next these (white peaches from Phila.) 25 Mississippi Nuts—something like the pig nut—but longer, thinner shelled, and fuller of meat." On May 2, 1786, Washington made a second planting. He writes, "planted 140 seed sent me by colo. Wm. Washington and said by him to be the seed of the large magnolia or Laurel of Carolina. . . . Also 21 of the Illinois Nuts; compleating [sic] at the No. end; the piece of a Row in my Botanical Garden in which on the day of I put Gloucester hickory Nuts." Again in 1794 Washington noted the planting of "several Poccon or Illinois nuts" around Mount Vernon.[51]

The first president enjoyed not only growing pecan trees; he was known to be quite fond of the pecan's taste as well. In a widely reported account, a Frenchman by the name of DeCourset, who had served with Washington during the Revolutionary War, remarked in his own journal that "the celebrated General always had these nuts and was constantly eating them."[52]

George Washington was not the only founding father to be smitten with the pecan. Thomas Jefferson, perhaps more so than Washington, had much to do with popularizing the pecan in early Colonial America. Thanks to Jefferson's detailed notes on his farm, Monticello, and his voluminous correspondence, we have a very good record of his association with pecans.

Jefferson's first mention of pecans is found in his *Notes on the State of Virginia*, written between 1781 and 1785. There he lists the "Paccan" or "Illinois nut" growing on the Illinois, Wabash, Ohio, and Mississippi Rivers. Shortly thereafter, as ambassador to France in 1786, Jefferson requested that Francis Hopkins of Philadelphia send him some pecans: "Dear Sir . . . procure me two or three hundred Paccan nuts from the Western country. I expect they can always be got at Pittsburgh, and am in hopes that by yourself or your friends some attentive person there may be engaged to send them to you. They should come as fresh as possible, and come best I believe in a box of sand. Of this Bartram could best advise you. I imagine vessels are always coming from Philadelphia to France."[53]

Upon returning from France in 1790, Jefferson made arrangements to have pecan nuts planted at Monticello in December of that year. He wrote his friend Thomas Mann Randolph, "I send herewith some seeds which I must trouble you with the care of. They are the seeds of the sugar maple and the Paccan nuts. Be so good as to make George prepare a nursery in a proper place and to plant in proper season. Mr. Lewis must be so good as to have it so inclosed [*sic*] as to keep the horses out." The nuts are believed to have been obtained from the Ohio or upper Mississippi River Valleys.[54] Randolph reported back to Jefferson on the status of the pecans in July 1791: "The pacans have not appeared as yet. . . . Thinking that they would not bear transportation I took the liberty to place them partly on each side of the new way leading from the gate to the house & partly in the garden. Several of those in the garden were destroyed unluckily by the hogs before it was enclosed."

Jefferson records the planting of 200 more pecan nuts in March 1794, and again in 1799, 1806, 1807, and 1808. Thomas Jefferson saw early on the potential of planting pecan groves. In January 1800 he wrote to Daniel Clark Jr. of New Orleans regarding the pecan tree's ability to grow in Virginia: "Two young trees planted in that part of the country in 1780, and now flourishing, though not bearing, prove they may be raised there; and I shall set great value on the chance of having a grove of them." Upon sending Thomas Lomax some young pecan tree seedlings in 1809, Jefferson remarked, "I propose to make a large orchard of Paccan." Although nothing significant was to come of Jefferson's pecan plantings, his enthusiasm for the tree and its fruit no doubt drew the attention of others to the pecan.[55]

Another early proponent of pecan production was André Michaux, an early-nineteenth-century French explorer. Michaux was nearly 30 years old when he took up the study of botany only after the death of his wife. In 1782 he was dispatched by the French government on a botanical expedition to Persia, where he was robbed of nearly everything but his books. Appointed royal botanist by Louis XVI, Michaux was then sent to the United States in 1785 to identify plants that could be useful to the French.

When he arrived in Kentucky in the 1790s, it was rumored that Michaux and his fellow French scientists had been sent to the western frontier as agents to enlist troops and to receive supplies for an attack on New Orleans. In the last decade of the eighteenth century, scientific explorations were often mixed with political intrigue. In 1792, then secretary of state, Thomas Jefferson persuaded the American Philosophical Society in Philadelphia to establish a fund to finance an expedition to locate the Northwest Passage. Jefferson handpicked Michaux to head this expedition. At the time, the Mississippi River was a vital trade route for the western states and territories, yet it was still under the control of the Spanish at New Orleans. When President Washington learned of the French activities and of André Michaux's role, he began efforts to quell this "conspiracy." Washington also cancelled Michaux's mission to locate the Northwest Passage, an expedition that would be fated to wait another 12 years, with Meriwether Lewis and James Clark at its head.[56]

Michaux first encountered the pecan around 1794 near Louisville, Kentucky. In 1819, after seeing stands of wild pecans managed by Indians near Kaskaskia, Illinois, Michaux was convinced that the pecan nut was worthy of cultivation because it was such a beautiful tree, and one that would produce excellent nuts. He spoke of the pecan as being "more delicately flavored" than the walnuts of Europe. Michaux believed the pecan would likely be more rapidly adapted to commercial production in the east if it were grafted onto black walnut, which he presumed would speed up the tree's growth.[57] Although it would not be through the use of black walnut as a rootstock, grafting would prove to be the keystone that would make pecans profitable.

· 2 ·

The Legacy of Antoine

The old Mississippi River Road in southern Louisiana is lined with mile after mile of relics of the Old South. The 70-mile corridor dripping with Spanish moss hugs both sides of the Mississippi River just southwest of Lake Pontchartrain and is home to some of the finest examples of Greek Revival architecture found anywhere. Mark Twain wrote of the Great River Road in *Life on the Mississippi*, "From Baton Rouge to New Orleans, the great sugar plantations border both sides of the river all the way, . . . Plenty of dwellings . . . standing so close together, for long distances, that the broad river lying between two rows, becomes a sort of spacious street. A most home-like and happy-looking region."[1]

The proximity of New Orleans to the River Road's plantations and their early role in the development of pecan production was no accident. The lavish estates of the River Road provided ample land, cash, and individuals committed to working the land and who enjoyed tinkering with Mother Nature. New Orleans had a major role in the early development of the pecan as a cash crop because it provided a large market for the crop and for planting stock. The Crescent City served as a center from which wild, seedling, and improved pecans were redistributed into Virginia, Pennsylvania, New York, and the non-native (to the pecan tree) southeastern regions of the country.

Nineteenth-century sugar planters built most of the large plantation houses along the River Road about 30 years prior to the Civil War. While they are often presented as Southern versions of utopia for the planters, the plantations were factories with the primary goal of producing a cash crop, in this case sugar, for world export. In the 1800s, a few of the major houses were lost to the ravages of the Civil War in one form or another. Here, Twain describes the appearance of the plantations in the late 1800s following the war: "All the procession paint the attractive picture in the same way. The descriptions of fifty years ago do not need to have a word changed in order to exactly describe the same region as it appears today—except as to the 'trigness' of the houses. The whitewash is gone from the negro cabins now; and many, possibly most, of the big mansions, once so shining white, have worn out their paint and have a decayed, neglected look. It is the blight of the war. Twenty-one years ago everything was trim and trig and bright along the 'coast,' just as it had been in 1827, as described by those tourists."[2]

In the 1920s, mosaic disease ravaged Louisiana's sugarcane crop,[3] and as a result, the sugar industry itself, leaving many of the grand mansions abandoned and in ruin. In the 1940s, restoration of the River Road's plantations took place in earnest. Even here, on River Road, just as in many areas of the Louisiana coast, the presence of big oil can be found. One of the road's most grand plantation mansions, San Francisco Plantation house, was restored with much financial assistance from Marathon Oil and is now said to have one of the finest antique collections in the nation. Today, tourists from New Orleans can see the grand Mississippi River Road and its plantation mansions through any number of tour companies operating out of the Crescent City. As they travel the old avenue of Southern aristocracy, they find the grand plantations interspersed with petrochemical plants and suburban strip developments in stark contrast.

Despite the upheaval of the sugar industry, one plantation, Oak Alley, with a significant tie to pecans, became the first River Road plantation to undergo restoration in the 1920s, under the guidance of owners Mr. and Mrs. Andrew Stewart. What is today known as Oak Alley Plantation was built in 1837 to 1839 by George Swainey for Jacques Telesphore Roman. Originally named Bon Sejour, Oak Alley is most recognizable by its plantation house, adorned with 28 enormous Doric columns, designed by Roman's father-in-law, Joseph Pilie. The house itself lies at the end of a double row of mammoth-sized live oak trees, from which the plantation derives its name.

It was here, on Oak Alley Plantation, that pecan production as it is known today is said to have been born. While people had been harvesting and consuming pecans for centuries, and even selling the nuts from groves of native pecan trees, pecan production as an industry did not seem to be a viable option. The tree's production was too sporadic, and the variability between the nuts produced by one tree and the next, too great. This all changed thanks to the skill of a slave at Oak Alley in the winter of 1846.

While in many circles, the term "seedling" refers to a young tree, to those familiar with pecans, a "seedling" refers to a tree grown from a nut. Although many cultivars have arisen as chance seedlings, most seedlings are, in a sense, wild pecans arising from a female flower that has been pollinated and fertilized by the pollen of a male flower from another tree. No two pecans are therefore alike. One could plant 1,000 nuts from the same tree and most of them would grow into trees that produced pecans with little resemblance to their parents or siblings. As a result, pecans are found in a plethora of sizes, shapes, shell thickness, and kernel quality. The tree itself may also vary considerably from its relatives in its habit of growth, foliage density, leaf shape, time of budbreak, and so on. This puzzled many of the early pioneers of pecan culture, who scratched their heads in wonderment at the thought that nuts from a favored tree would not grow into a tree resembling the original. The first decade of the twentieth century had passed before most people working with pecans recognized this phenomenon as fact.

Most cultivated plants grown today, including pecans, are horticulturally improved forms that owe their continued existence to the fact that humans propagate them. Selected individual trees are asexually propagated through grafting or budding. These terms refer to methods of connecting two pieces of living plant tissue together so that they will unite and then grow and develop as a single plant. This allows shoots or buds (termed scions or graftwood) from a tree with desirable characteristics to be transferred or attached to an established tree called the rootstock. Those who practice grafting consider it as much an art as a science. Over the centuries, grafting has always held a rather mystical aura. Through the Middle Ages, grafting was considered a secret by those in the know and a miracle by the uninitiated. Even as late as the early twentieth century, legendary horticulturist Liberty Hyde Bailey wrote that many horticultural writers considered grafting "akin to magic and entirely opposed to the laws of nature."[4]

Although it is difficult to pinpoint the date of origin for grafting, this process of joining two plants is an ancient practice. Though potential references to grafting exist in the Hebrew Bible, they are only inferred references at best. The Mishna, the first redaction of the Jewish oral traditions, dates to 220 BC and contains numerous references to grafting. Such references can be found in the Christian Bible as well, where the apostle Paul uses the practice of grafting olive trees to illustrate his point in Romans 11:16–24.

The ancient Greeks are known to have had a working knowledge of grafting. The earliest verifiable written account of grafting is from the Hippocratic treatise "On the Nature of the Child," written about 424 BC by various followers of Hippocrates. It states, "Some trees however, grow from grafts implanted into other trees: they live independently on these, and the fruit which they bear is different from that of the tree on which they are grafted. This is how: first of all the graft produces buds, for initially it still contains nutriment from its parent tree and only subsequently from the tree in which it was engrafted. Then, when it buds, it puts forth slender roots in the tree, and feeds initially on the moisture actually in the tree on which it is engrafted. Then in course of time it extends its roots directly into the earth, through the tree on which it was engrafted: thereafter it uses the moisture which it draws up from the ground." Though not entirely accurate, the depth of understanding conveyed in this passage suggests a good working knowledge of grafting, an apparently common practice that must have been centuries old by that time. Aristotle wrote, "Grafting of one on another is better in the case of trees which are similar and have the same proportions." His student, Theophrastus, often referred to as the father of botany, discussed grafting techniques and the importance of avoiding desiccation or drying out of the grafting material. Likewise, the earliest of Roman writers, Cato, described various methods of grafting that are still used today, as did the author of the *Aenid*, Virgil, and Pliny the Elder.[5]

Although it cannot be documented with certainty because of the loss of historical records, some suggest that grafting was first detailed by the Chinese as early as 5000 BC, when Feng Li, a Chinese diplomat, became so enamored with the process of grafting peaches, almonds, persimmons, pears, and apples that he gave up his diplomatic career to pursue fruit production; however, questions remain regarding this assertion. Others claim that grafting may have originated in China in the second millennium BC. Regardless of where or when it began, grafting is an ancient art that has become most useful for pecan production.

The use of grafting and budding in pecan propagation imparts another level of complexity to its culture. Most tree crops are propagated in the same manner as pecan, that is, a piece of scion, either graftwood or budwood, is joined to a rootstock. Although obscurely understood, rootstock influences can include such important performance characteristics as dwarfing, precocity, salt tolerance, nutrient uptake, and resistance to cold, insects, disease, and nematodes. The influence of rootstock on plant growth and production has been observed for centuries. For example, in sixth-century China, Jia Sixie wrote that pear grafted onto "tang" would produce large pears with fine flesh, as compared to pear grafted onto "du," which yielded pears of poorer quality.[6] As a result of such disparity, fruit and nut nursery workers today may choose clonal rootstock for some crops. Clonal rootstocks are genetically identical and exhibit uniform behavior when planted in similar situations.

Source trees for clonal rootstocks are used from which cuttings are taken to grow the rootstock. For example, EMLA III apple rootstock provides dwarfing and resistance to woolly aphids and collar rot. Duke 7 avocado rootstock provides tolerance to root rot, as does RX 1 walnut rootstock. Seedling rootstock, on the other hand, refers simply to rootstock trees grown from seeds that are usually open pollinated, or in other words, pollinated by pollen that chances to blow by in the wind. Just as in nut production, each one of thousands of seeds planted from the same tree will result in unique trees with their own characteristics. Usually this performance varies based on geographic origin of the seedstock families. With regard to pecan, nursery workers in different regions today often use different rootstocks. In the southeastern United States, the most commonly used rootstocks are seedlings grown from Elliott seeds. Their use extends into East Texas, but because growing conditions vary across Texas, multiple rootstocks are used throughout that state. In North Texas, Giles is popular. Riverside and VC-168 seedstock are commonly used in the arid lands of the Western United States because of their salt tolerance and vigor, respectively. The most vigorous rootstocks come from trees originating in the southern portion of the pecan's range, while those from the north are usually hardier.

Interestingly, you don't always have to graft onto rootstock of the same species. Almonds are grown on almond rootstock but also on peach, peach-almond, and plum. Likewise, peach can be grown on almond or plum. Pecans can be grafted onto other hickories. Grafting pecan onto water hickory (*Carya aquatica*) can enhance the tree's survival on wet, poorly drained soils. However, because most pecan orchards are planted on well-drained upland sites, pecan nurseries generally use seedling pecan trees as rootstock.[7] Regardless of the rootstock used, grafting or budding is ultimately essential for uniform nut production. The major difference between grafting and budding is that with budding, only a single bud is united to the rootstock, whereas with grafting, a larger piece of wood containing several buds is used.

Attempts at budding pecan occurred as far back as 1822, when Dr. Abner Landrum budded pecan onto hickory rootstock at his plantation in Edgefield, South Carolina. Dr. Landrum was a fascinating man with a multitude of interests. Laboring to improve the way of life for those in South Carolina, Landrum is credited with unlocking an ancient Chinese method of producing stoneware pottery. A lead glaze was used on most of the earthenware pottery produced in the South during the 1700s and early 1800s. But because people ate from these plates and preserved foodstuffs in large containers made by this method, lead poisoning became a common occurrence. Acids from the vinegar used in preserving seemed to accelerate the process. A highly enlightened and educated man, Dr. Landrum read extensively from his vast library in attempting to develop a better method of producing safe stoneware. Finally, Dr. Landrum and his brothers developed a method using local materials, including a mixture of the kaolin clay found in the area along with sand and ashes. By this method, Landrum was able to produce the first alkaline-glazed stoneware pottery in the New World, combining the techniques of Europe and Asia and creating a viable alternative to the lead-leaching dishes. Landrum's pottery would spread throughout the entire southern tier of states to Texas during the late 1800s, forestalling countless agonizing deaths as a result.[8]

By the end of the first decade of the nineteenth century, Abner Landrum was manufacturing his stoneware in large quantities on his property, which came to be known as Landrumsville or Pottersville. One of Dr. Landrum's slaves, known only as Dave, would, with Landrum's help, become a literate, skilled craftsman, whose proficient pottery skills would make him famous.[9] Dave often inscribed short poems onto his vessels, which are now world famous. Most today fetch a handsome price and can be found residing in prominent collections and museums throughout the world.

Dr. Landrum also became a publisher, printing his periodical, *The Hive*, a Unionist-slanted publication with articles on science and the arts. Because of his many interests and activities, Abner Landrum's horticultural pursuits are often overshadowed. Perhaps for this reason, his contribution to the pecan industry has been largely overlooked. Discussing his attempts at budding pecan in an article for the *American Farmer*, Landrum wrote, "The pecan did not appear to take so well as the walnut but my trials were made rather later in the season." His second attempt appears to have been more successful, as he states, "I have this summer budded some dozens of pecan on the common hickory nut, without a single failure as yet; and some of them are growing finely."[10] For whatever reason, Landrum's attempts at budding pecan never caught on with other nurserymen. There were apparently no viable markets nearby to support such efforts. Landrum's propagation never led to any significant orchard establishment or nursery sales, and as a result, his contribution to the pecan industry has been largely ignored.

Both grafting and budding require considerable skill, the mastery of which comes only with experience. Necessitating much practice to perfect, budding and grafting both take advantage of the process of cell division within the plant. All plants grow by cell division in three areas—the shoot apex (tip), the root apex, and the cambium (the layer under the outer bark).

When plants are wounded, new cells are formed as callus tissue. As the freshly cut graftwood is placed in contact with the rootstock, the water- and nutrient-conducting tissues, or cambium, regions of each make contact. If the union between the rootstock and graftwood is secure, the outer layers of the cambium region produce cells called callus tissue, or parenchyma. Certain cells of this newly formed tissue in line with the cambium form new cambium cells, creating a bridge between the rootstock and graftwood by which water and food, primarily sugar, can move from the rootstock to the graftwood or budwood. The introduction and success of this basic technique to pecan culture would forever change the way humanity considered the tree and its nut.

The slave gardener of Oak Alley Plantation is known today only as Antoine. We have no record of his background and no details of his life, other than that he was owned by J. T. Roman and was mightily endowed with a green thumb. Antoine developed a reputation with plants, including a deft hand at the propagation of pecan trees. Records of Roman's estate show that he owned 113 slaves, 93 of which labored in the fields, while the remaining 20 worked in Roman's household. Among the field slaves, Antoine was listed as 38 years old, valued at $1,000, $500 less than Prince, a mulatto carpenter. The notations beside Antoine's name state that he was "a Creole Negro gardener and expert grafter of pecan trees."[11]

Sometime in the early 1840s, Dr. A. E. Colomb attempted to graft a favorite tree growing on the Anita Plantation of Mr. Amant Bourgeois, on the east bank of the Mississippi River in Saint James Parish. It is not known whether Dr. Colomb's tree was a chance seedling or a planted nut, but it regularly produced large, thin-shelled pecans. Dr. Colomb's efforts at grafting the tree failed, so he cut graftwood from the tree and took it to J. T. Roman and his talented gardener, Antoine, across the river at Oak Alley.

Choosing a site near the great mansion, Antoine set about grafting the trees. No one knows how many total trees Antoine grafted, but 16 of his grafts were successful. After this initial success, Antoine continued grafting trees on Oak Alley until 110 pecan trees were grafted in a large pasture located about 40 "arpents," or 7,680 feet, from the river, before the unexpected early death of J. T. Roman in 1848. By the end of the Civil War, 126 of Antoine's grafted trees were bearing at Oak Alley.[12] Following the war, Oak Alley was sold to John Armstrong and a quick succession of other owners, who promptly cut down many of the trees at the peak of their productivity in order to plant sugarcane. At the time, the nuts from these trees were reportedly selling at from $50 to $75 per barrel. By 1876, Oak Alley had fallen into the hands of a man named Hubert Bonzano.

In 1876, the 100th anniversary of the signing of the Declaration of Independence, Philadelphia was hosting the Centennial Exposition. The Centennial Exposition was the first official World's Fair held in the United States. Named the International Exhibition of Arts, Manufactures and Products of the Soil and Mine, the exhibition was held in Fairmount Park on the banks of the Schuylkill River and was attended by over 10 million visitors, roughly 20 percent of the population of the United States at the time.

The Centennial Exhibition was America's coming-out party. The opening ceremony culminated with President Ulysses Grant and Brazilian Emperor Dom Pedro flipping the switch on the newly developed Corliss steam engine, which powered most of the other machines at the exhibition. Other products first displayed at the Centennial Exhibition were Alexander Graham Bell's telephone, the Remington typewriter, Heinz ketchup, and the Wallace-Farmer electric dynamo, a precursor to electric lighting. In addition, the right arm and torch of the as-yet-to-be completed Statue of Liberty was displayed. By contributing a fee of 50 cents, attendees could climb a ladder to the balcony surrounding the torch. The fees were used to fund the rest of the statue.

Here, at the Centennial Exhibition, amid Bell's telephone, the Remington typewriter, and Heinz ketchup, Hubert Bonzano introduced the world to the first vegetatively propagated pecan cultivar. Horticultural Hall was unlike most of the buildings constructed for the exposition because it was meant to be a permanent structure. Designed by Hermann J. Schwarzmann, an engineer for the Fairmount Park Commission, who incidentally had never before designed a building, Horticultural Hall was styled in a Moorish design, with an iron and glass frame on a brick and marble foundation. It was here that Bonzano displayed nuts from the trees grafted by Antoine at Oak Alley Plantation. One of the nation's most notable scientists and botanists, Professor William H. Brewer, chair of agriculture at Yale's Sheffield Scientific School, awarded Bonzano a certificate for his pecans, commending their "remarkably large size, tenderness of shell, and very special excellence."[13]

Either here at the exhibition or sometime afterward, the variety was given the name Centennial, making it the first recognized pecan variety. The variety was first cataloged under the name Centennial in 1885 by Richard Frotscher of New Orleans, who in 1882, along with his partner in the pecan nursery business, William Nelson, was the first to offer grafted or budded trees for sale, including Centennial, along with Frotscher and Rome.

Centennial, like many recognized pecan varieties today, arose as a chance seedling, most likely planted by either human or beast. Regardless, it was selected for its superior qualities, just as many subsequent trees have been, and was then vegetatively propagated. Centennial became the first pecan variety planted in the form of a commercial orchard, with the idea of producing nuts for sale.

The original or "mother" Centennial tree, from which the first graftwood was taken by Dr. Colomb in the 1840s at Anita Plantation, was destroyed on March 14, 1890, when it was lost to the Anita crevasse, which swept away the earth beneath the tree to a depth of 15 feet after a defective rice flume (used to obtain water for flooding the rice fields) caused a breach in the levee. Today

FIGURE 1 One of
only a few remaining
Centennial pecan trees
photographed in 1941.
The tree was origi-
nally grafted by a slave
known as Antoine on
Oak Alley Plantation
in Louisiana.

the crevasse appears as just another low, swampy depression along the Mississippi River. Of the first trees grafted by Antoine on Oak Alley Plantation,
only two were still alive in 1902.[14]

Centennial is not widely cultivated in today's commercial pecan orchards,
largely as a result of the amount of time it required—as many as 15 years from a
bud or graft—to begin bearing pecans under turn-of-the-century pecan management practices. This "handicap" led to the removal of Centennial from most
recommended planting lists as early as 1906. Many varieties planted in today's
orchards will begin bearing nuts in the third or fourth year, with most requiring only six to eight years for a commercial harvest. Modern observations of

Centennial have demonstrated that under today's management, the variety's precocity is not much different from that of other popular varieties. But, compared with that of today's varieties, the yield capacity of Centennial is low, preventing most growers from considering it as a potential choice for commercial production.

The real significance of the Centennial story is that Antoine's successful grafting technique, for the first time, allowed those interested in growing pecans to select the best seedling trees and asexually propagate them. Grafting allowed for the production of trees with reliably predictable characteristics. This major technical breakthrough was slow to be adopted, however, and had little commercial impact until the late 1800s, when nurserymen in Louisiana, Texas, and Mississippi used the technique on a commercial scale. This advance led to a profusion of new seedling selections around the turn of the twentieth century, which were propagated and sold for commercial planting.

The years immediately following the Civil War saw poverty and strife ravage the southern United States. The farm economy of the former Confederacy was reeling from the aftermath of the war. Agriculture was still the biggest business in the South and to a large extent remains so today. But the late 1800s found many Southern farmers searching for alternative crops that could reduce their reliance on king cotton. Out of this economic environment the pecan evolved from what has been characterized as simply a salvage crop used mainly for home consumption or as a trade item into a legitimate orchard crop industry.

Were it not for Emil Bourgeois and E. E. Risien, who rediscovered Antoine's forgotten grafting techniques, the pecan industry would not enjoy the success that it does today. In 1836, Mr. Duminie Mire of Saint James Parish, Louisiana, dug a few holes in his garden and planted a handful of pecan nuts that he had collected from a "highly esteemed tree" belonging to his neighbor, a Mr. Gravois. A few of these nuts germinated and grew into pecan trees, but one in particular proved to be far superior to the others. The tree consistently produced nuts that were long and pointed at one end, with creamy, golden kernels. According to Mr. Mire, the pecans produced from his tree were very similar to the nuts he had planted from the parent tree. As the tree's reputation spread, it attracted the attention of Mr. Emil Bourgeois.

Emil Bourgeois was born in 1832 in Saint James Parish, only a few years before Duminie Mire planted the nuts in his garden. A large man with dark complexion, dark hair, and hazel eyes, Bourgeois served as a private in Company A of the 4th Louisiana Cavalry during the Civil War and was captured near New River on December 17, 1864. Following the war, he returned to Saint James Parish and became a man of some renown in central Louisiana. Bourgeois was considered a cultured gentleman and enjoyed the privileged life of a nineteenth-century sugar planter. He had been educated in Kentucky, served on the Pontchartrain levee board, and had made his farm, Rapidan Plantation, one of the best sugarcane-producing plantations in the state.[15]

In 1877, Bourgeois removed 22 shoots from Duminie Mire's tree and grafted them onto seedling trees at Rapidan Plantation, reviving the techniques of Antoine. Exactly half of the grafts were successful. When the trees began producing nuts, Bourgeois began propagating young trees for orchard planting and sold them to his neighbors. The tree was known locally as the Duminie Mire pecan. It was later renamed Van Deman by Colonel W. R. Stuart of Ocean Springs, Mississippi, in honor of Professor H. E. Van Deman, a respected USDA fruit scientist. The Van Deman became widely advertised in the 1890s and, being one of the first grafted pecan trees offered for sale, it was distributed throughout the pecan-growing regions of the South. By 1912, it was one of the most widely disseminated pecan varieties.[16]

The summer of 1872 found 19-year-old Edmond E. Risien arriving in the port of Galveston, Texas, from his home in Kent, England. After taking in the sites of Galveston, he made his way to Limestone County in east-central Texas, where he planned to visit relatives. Risien had in mind to travel on to California after his time with family; however, his plans were altered. Passing through the town of San Saba, he decided to stop and make a little extra money. Risien was a cabinetmaker by trade and found himself with the unexpected opportunity to craft a set of coffins for three men who had recently been hanged. The young cabinetmaker liked what he saw in the little town and decided to set up shop and stay for a while.

E. E. Risien had a wide range of interests. Aside from his cabinetry business, Risien kept bees and would later design and operate San Saba's first waterworks and build the area's first hospital.[17] In those early years of Risien's arrival to the Texas Hill Country, wagons loaded down with buffalo meat and pecans filled the streets of San Saba each fall. Most of the land surrounding San Saba was unoccupied, and locals often gathered pecans from the trees growing along the nearby San Saba and Colorado Rivers.[18] Something about the sweet nuts took hold of Risien's curious mind. He became fascinated with pecans and offered a five-dollar prize for the best nut brought into his store. As the wagonloads of pecans flowed in, Risien judged each nut on its color, flavor, texture, and ease of shelling.

One pecan in particular got Risien's attention. It was small but had exceptionally well-filled, kernels of a light golden brown color. After inquiring about the tree that produced this nut, Risien was led to a peninsula at the junction of the San Saba and Colorado Rivers. There at the confluence of the two rivers was a forlorn tree with only a single branch. Shocked, Risien asked his guide what had happened to the tree. The man replied that he had left the one limb so that he could stand on it while sawing off the others to get to the pecans.[19] This was not an uncommon practice for harvesting nuts in the area at the time and had previously been used by Indians and settlers throughout the pecan's range.

Risien, however, was not deterred. He was so impressed with the nut that he bought the land and the surrounding 320 acres from the absentee landowner. Slowly, the tree grew a new canopy and once again produced Risien's prized nuts. He named the tree the San Saba and it was from this tree that Risien began the first known pecan breeding program, somewhat by accident.

Risien set about taking pollen from a certain tree on the property and pollinating the female flowers of the San Saba tree by hand in a painstaking process. He rode hundreds of miles on horseback in search of a suitable tree with which to cross the San Saba. He carefully placed the long male flowers, called catkins, into his saddlebags and transported them back to the Mother Tree for pollination. Risien finally planted 1,000 nuts on 30 acres of the land he purchased, expecting each nut to produce the same tree. He soon found out that pecans do not produce true to type, meaning that each offspring can exhibit dramatically different characteristics from those of its parents. Although Risien may have originally considered his planting a failure, it turned out to be a revelation. Risien experimented widely with various methods of propagation, but his most successful work hinged on a technique termed "ring-budding," in which a double-bladed knife was used to make precise ringed cuts of scion and rootstock, which matched perfectly. This method proved easier and more reliable than previous methods of propagation.

Risien founded the West Texas Pecan Nursery in 1888 along the tree's native river bottoms. Ten years after planting the nuts, he took graftwood from promising selections of his 1,000 nut plantings and grafted them onto older trees. From his failed experiment, Risien developed the San Saba Improved, Texas Prolific, Onliwon, Squirrel's Delight, No. 60, and what is still today the most dominant pecan variety grown in the western United States and Mexico, Western Schley. Running a telephone line from his nursery to the post office in nearby Rescue, Texas, Risien was able to prepare orders received in the afternoon mail for next-day shipment. He developed a catalog, showcasing both his plants and his personality to spur interest. He also shipped barrels of pecans to children in his English hometown and to soldiers overseas.

Risien's work made him a legend in the nursery business and in the fledgling pecan industry during the twentieth century. He shipped nuts all over the United States, and eventually the world. His customers included Queen Victoria, President William McKinley, and Secretary of State John Jay. Alfred Lord Tennyson, England's poet laureate and author of the poetic tribute to British cavalrymen "The Charge of the Light Brigade," was a fan of Risien pecans. On the occasion of Risien's 39th birthday, the poet wrote him, wishing him a "long and happy life to see his pecan trees flourish." Also impressed with Risien's pecans, C. W. Post derived the name Grape-Nuts for a cereal from Risien's exhibit of a large cluster of pecans at the Chicago World's Fair.

E. E. Risien corresponded regularly with the man whom many called a botanical wizard, Luther Burbank. Creator of the Russet Burbank potato (found in the majority of McDonald's french fries), the Shasta daisy, and the Santa Rosa plum, Burbank had the greenest thumb in the country during the late nineteenth and early twentieth centuries, even breeding the fuzz off of peaches. Shunned by many academics as "nonscientific" because of his failure to keep careful records, Burbank developed more than 800 plant varieties on his California farm over his 55-year career. Burbank was more interested in results and in helping others than in pleasing academics, and by most accounts he did just that. His accomplishments later gained him the respect of many of his critics in academia who could no longer scoff at the results he achieved.

Burbank and Risien struck up a correspondence, interested as they both were in the production of plants. After observing Risien's work, Burbank once commented, "If I were a young man I would go to Texas, knowing as I do the possibilities of the pecan industry, and devote my life in propagating new species of pecans and doing the same work there in nut culture as I have done in other lines of horticulture. Your pecan is superior to our walnut and you are standing in your own light. Why not develop it? I cannot think of any kind of diversification likely to pay the southern farmer as well as pecan growing."[20]

If Oak Alley Plantation was the birthplace of commercial pecan production, then Jackson County, Mississippi, was its cradle of infancy. The Gulf Coast area surrounding Ocean Springs, Pascagoula, and Scranton became an enclave for foreign immigrants following Mississippi's statehood in 1817. Seamen and laborers flooded into the area from France, Spain, Portugal, Germany, and Ireland. The immigrants set about building the new lives they had dreamed of through shrimping, fishing, farming, timbering, and naval stores.[21]

One of the nation's oldest cities, Ocean Springs was founded in 1699 by Pierre Le Moyne d'Iberville under the authority of King Louis XIV. Originally called Fort Maurepas, it was the first permanent French outpost in French Louisiana and was established as a foothold to prevent Spanish encroachment on France's colonial claims. The site was maintained well into the early eighteenth century. From this area sprang individuals gifted with a curiosity and ability for growing things.[22]

During the 1920s there were 33 pecan nurseries in Jackson County, Mississippi, which is probably more than the number of pecan nurseries in the United States today. This small area in the southeastern corner of Mississippi was a harbor of diversity, not only in the ethnic background of its inhabitants, but in its variations of pecan as well. The region was particularly well suited to the early development of the pecan industry for a number of reasons. Although lying just outside the native range of the pecan, Jackson County was close enough that an ample diversity of nuts and seedling trees could be easily obtained. Jackson County was also close to New Orleans, considered the major center of pecan trading in these early years. Perhaps most importantly, the region was home to visionary men of industrious minds with competent skills in agriculture and horticultural techniques who recognized the potential of the pecan. From among the incredible pecan diversity that arose here sprang four varieties—Stuart, Schley, Pabst, and Alley—known in pecan circles as "the big four" that would come to dominate the southern pecan industry from the 1870s until World War II.

In the 1870s several farmers began planting and developing orchard crops, primarily citrus and pecan, in the Jackson County area. One of the first of these was Colonel William Rasin Stuart.

A retired sugar and cotton broker from New Orleans who moved to Ocean Springs at the age of 56, Stuart had a keen interest in agricultural pursuits and came to be considered an expert on topics ranging from sheep to pecans. Around 1875, Colonel Stuart bought 100 seedling pecan trees from Mobile and New Orleans to plant an orchard on his farm at Ocean Springs, with the idea of growing pecan seedlings for sale. He established the Stuart Pecan Company, and in an attempt to develop a profitable pecan variety, Colonel Stuart "planted the largest and best flavored pecans that could be found, without regard to price."[23]

Only a year before Colonel Stuart planted his orchard, John R. Lassabe of nearby Pascagoula established an orchard from nuts brought from Mobile,

FIGURE 2 Colonel
W. R. Stuart, "Father
of the Paper-Shell
Pecan" and namesake
of the Stuart pecan, one
of the most popular and
widely planted pecan
varieties in the world.
From a pamphlet dis-
tributed in 1893 by the
now-defunct Stuart
Pecan Company.

Alabama. The orchard was later purchased by Captain Eugene Castanera, post-master at Moss Point, Mississippi. One tree from this orchard soon developed a reputation locally as a fine bearer of large, well-filled, good-quality pecans. The average yield of the original tree from year 15 to 18 was about 140 pounds per year. In the last of those years it produced a whopping 350 pounds of nuts, which were sold at one dollar per pound, a remarkable feat for a tree under 20 years old. Although originally called the Castanera pecan in that region of southeastern Mississippi, it was grafted in 1886 by A. G. Delmas of Scranton, Mississippi, who took over 60 pieces of graftwood from the original tree to propagate. Of all the grafts he made, only one survived. The Castanera was later propagated around 1890 by John Keller, a partner of Colonel Stuart in the nursery business. Unaware that the tree had previously been called Castanera, H. E. Van Deman, the USDA pomologist for whom the Van Deman variety was named, gave it the official name of Stuart, after Colonel W. R. Stuart.[24]

The Stuart pecan quickly became the most popular and widely planted pecan variety in existence. The tree was resistant to winter injury and broke bud late enough in the spring to avoid late freezes, which allowed it to be planted over a wide geographical area from Texas to Georgia. It was resistant to disease, and its large size made Stuart highly sought after for the in-shell pecan trade. Stuart was highly promoted by pecan nurserymen, including Colonel Stuart himself. Because no other pecan variety could produce so many nuts of such good quality, Stuart remained the most widely planted pecan variety in the world for over half a century. Today it is still the most abundant tree in pecan orchards of the southeastern United States, and it accounts for over one-fourth of the trees in the improved-variety orchards of the world.

Although Colonel Stuart was there at the beginning of the propagation and dissemination of the Stuart variety and no doubt played a large role in its promotion, the full credit for this variety's perpetuation should be shared with John Lassabe, Eugene Castanera, A. G. Delmas, and John Keller. Still, the Stuart pecan was not Colonel Stuart's only contribution to the development of the industry. Upon his return from a trip to Georgia, Colonel Stuart shipped a large volume of Stuart pecans by rail to San Francisco in November 1890, where they were loaded on board a steamship headed for Melbourne, Australia.[25] This was most likely the first time pecans made their way to the Land Down Under, which would one day have a thriving pecan industry of its own.

Called the "Father of the Paper-Shell Pecan," Colonel Stuart was quoted in the *Atlanta Constitution* in 1890 as saying, "I began [to cultivate pecan] at fifty-six years of age. I am now seventy-one, and make more money out of pecans than I do out of cotton. The young man in the South ought to think of this. There is unlimited money in pecan culture in the South, and I am anxious to see our people plant pecan trees just as they do apple or peach trees. They will make the South rich." Reading a paper before the Mississippi State Horticultural Society on December 11, 1891, Colonel Stuart further espoused his faith in the pecan by the remark, "I am now in the 72nd year of my age, and as an evidence of my faith in this industry, will state that I have just finished clearing up a piece of new ground in which I will plant a young grove this winter." Colonel W. R. Stuart would not see those pecans bear fruit. He lived only another three years, but the name Stuart continues to be a prominent one anywhere pecans are grown.

Sometime in the mid-1870s, James Moore, a blacksmith in Ocean Springs, sold some pecan seeds to Colonel Stuart. Having more than enough seedlings to accomplish his purposes, Colonel Stuart sold five of them to a neighbor, Peter Madsen. A few years later, Madsen sold his property, including the five pecan seedlings, to Mrs. H. F. Russell. Of the five trees that grew on the property, four produced large, thin-shelled pecans. The best of these was propagated and later named Russell by Mississippi's first pecan nurseryman, Charles E. Pabst.

Charles Pabst was born in Schleswig-Holstein, Germany, in December 1850. Along with his friend Augustus von Rosambeau, Pabst left his homeland for greener pastures in 1866. An initial attempt to settle in Australia being unsuccessful, the two friends, like so many others, set their sights on the shores of America. Upon arrival, the two young Germans found work on the sugar plantation of Leon Godchaux, where von Rosambeau was employed as a sugar chemist while Pabst toiled as a sugar cooker. Later, the two moved to Ocean Springs under the employment of a wealthy New Orleans industrialist, Ambrose Maginnis. The pair were caught up in a peanut-growing venture with Maginnis, but that plan failed. Yet they had finally found their home in Ocean Springs.

While von Rosambeau eventually became a successful merchant, Pabst became involved in carpentry, helping to build the town's First Presbyterian Church, where he played the organ during services. Pabst bought three lots in town for $80 in 1881 and, two years later, started Charles E. Pabst & Sons, which would later become Ocean Springs Pecan Nursery.[26] This endeavor was the first of many pecan nurseries in Mississippi. On the 20-acre nursery, Pabst planted approximately 400,000 pecan trees, later adding another 40 acres to accommodate his business.

Around the same time that James Moore sold Colonel Stuart his pecan seeds, William B. Schmidt planted pecan nuts purchased in New Orleans to establish an 11-acre pecan orchard on his farm in Ocean Springs, making this eventual orchard one of the first, if not the first, in the area.[27] Although not a full-time Ocean Springs resident, William Schmidt was a great contributor to life in Jackson County. Officially a resident of New Orleans, Schmidt owned the Ocean Springs Hotel, an estate called Summer Hill on the front beach, and several other properties throughout the town of Ocean Springs.

William B. Schmidt was born in Schwenningen, Baden-Württemberg, Germany, on April 10, 1823. His parents immigrated to the United States and attempted to settle in Saint Louis, Missouri, and later, Lexington, Kentucky, before finding a home in New Orleans in 1838. At the age of 22, William Schmidt established a business partnership with Francis M. Ziegler. Their firm, Schmidt & Ziegler, began as a small wholesale grocery business. By the end of the nineteenth century, Schmidt & Ziegler had expanded to 11 stores and later established international trade with South and Central America. Described as a quiet, thoughtful man with a will of iron and a heart of gold, Schmidt gave generously to the people and town of Ocean Springs, over which he would have a lasting influence.

The *Mexican Gulf Coast Illustrated* reported in 1893:

The grounds of Mr. Schmidt are the largest and most highly improved in the place [Ocean Springs] and are among the finest on the Coast. Besides the improved grounds there is a park of several acres. The family residence is not pretentious but very comfortable and supplied with modern conveniences. It is lighted with gas manufactured on the place. An artesian fountain gives a full supply of water brought from a depth of 450 feet; a hydraulic ram forces the water over the place. There are several fish ponds supplied with green trout [bass] and other kinds of fish. The grounds immediately around the residence are highly improved and richly ornamented with rare flowers and plants. Mr. Schmidt is not only a very successful business man, but keeps abreast with the latest improvements in whatever pertains to matters where his interests are affected. He has adopted sub-irrigation and sub-drainage on his own extensive grounds where vegetables and fruits for his own use are grown, and his table in the Crescent City when the season for their maturity arrives, is supplied with these products fresh from his own grounds. He has the Scuppernong grape, the pecan in its best state, and other specialties too numerous to mention in detail. There are also a number of high bred milch cows kept, and every morning the Coast train takes fresh milk to the city [New Orleans] for the use of his family. The yield per acre of crops of vegetables raised is often phenomenal.[28]

Charles Pabst and William Schmidt's paths would eventually cross to make a contribution to the world of pecans. One of the original nuts planted by Schmidt would grow into a grand tree that would catch the attention of Pabst, who propagated the tree in his nursery in 1890. The tree produced consistent yields of good-quality nuts, and in 1893, B. M. Young of Morgan City, Louisiana, named this tree in honor of Mr. Pabst. The Pabst pecan tree is very vigorous, with long vertical fissures in the bark, large dark green leaves, and spreading branches that form an open canopy. The Pabst is probably one of the more majestic pecan trees in appearance, growing to enormous size, lending it the nickname "the mighty Pabst." In older mixed orchards of the Southeast, Pabst trees are usually the largest trees in the orchard. As one of the "big four" varieties established in early southeastern pecan orchards, Pabst is commonly found in older orchards of the region; however, because it is difficult to control disease in such large trees and because more profitable varieties are available, Pabst is rarely planted today.

The Alley pecan is probably the least known today of the "big four." In 1871, Colonel R. Seal of Mississippi City, Mississippi, gave some pecans from an unknown tree to Mrs. C. H. Alley of Scranton in Jackson County. Mrs. Alley planted a nut in a box in 1871. The nut sprouted, and under Mrs. Alley's care, a tree began to grow. A year later, Mrs. Alley planted the tree in her garden. About nine years later, the tree began to bear nuts and developed a reputation as a consistent producer of small but good-quality pecans. This distinction attracted the attention of Frank H. Lewis of Scranton, who first propagated Alley in 1896 and began its introduction. Three years later, Charles Pabst would also begin propagating Alley and, through his extensive nursery operation, would help bring this pecan onto the general market.

There are two main reasons why Alley is not found in many orchards today. For one, Alley produces a small nut, and about 75 nuts are required to make one pound. The greatest demand is for large pecans, or those with a count of about 50 nuts or fewer per pound. Still, the Alley shells out particularly well into good-quality kernel halves. One major downfall of Alley is that it is highly susceptible to the greatest enemy a southeastern pecan faces, a disease called pecan scab, which manifests in the form of little black specks that appear on the leaves and developing fruit. As the disease progresses, scab can completely cover the nuts, causing them to drop off before they reach maturity.[29]

In the 1920s no pecan variety could compare with Schley. It was held up as the standard of pecan quality, and in the southeastern United States, it is still a preferred nut for many local consumers. In 1881, Albert Grant Delmas, a Jackson County chancery clerk and nurseryman, planted a nut at his home in Scranton that was reported to be from the original Stuart tree in Pascagoula. When the tree Delmas planted reached 17 years old, he christened it the Schley after Admiral Winfield Scott Schley, commander of US naval forces who was credited with the destruction of the Spanish fleet fleeing the Cuban port of Santiago de Cuba, virtually ending the Spanish American War. The record of the Schley pecan as a seedling of Stuart has long been in question because of the general lack of obvious Stuart characteristics in the nut and in the tree itself. Some accounts report that Schley did not develop from a Stuart nut at all, but instead grew from a nut that came from the same tree in Pascagoula from which the nut that produced the first Stuart tree originated. Recently, with the use of molecular genetics, the account of Schley as a Stuart seedling was disproven.

In 1900, Delmas began propagating Schley by grafting the scions onto older trees in a process called top-working. That same year, the Schley won a bronze medal at the International Exposition in Paris. Four years later it also won the silver medal at the exposition in Saint Louis. The Schley's popularity soared after introduction into Florida nurseries in 1902 by nurseryman D. L. Pierson, who obtained scions from the original Schley tree that same year. Pierson marketed the tree as Admiral Schley from his business, Summit Nurseries, in Monticello, Florida.[30] A 1908 Summit Nurseries ad in the *American Fruit & Nut Journal* called the Admiral Schley "the Pecan of the Future," further boasting that "no grove is complete without this variety which combines more fine points than any other."

The Schley truly does combine many fine attributes. At 65 nuts per pound, the nut is not large but is still within a range that makes it profitable. The tree is attractive, with dense foliage, large spreading branches that dip toward the ground, and vigorous sprouts emanating from the trunk. At the time of its introduction, Schley's greatest attribute was its thin shell, which was largely responsible for the best kernel quality (about 60 percent) anyone had seen at that point. In fact, it was the Schley that helped coin the term "papershell" pecan.

Schley was extensively planted in northern Florida and southern Georgia in the early 1900s as a result of D. L. Pierson's efforts. One can deduce that many old Schley trees in this area had their origin at Summit Nurseries by the abundance of Delmas trees found in the same orchards. Delmas is a variety named, of course, for A. G. Delmas. It, too, was a seedling that grew from a nut planted in 1877 in Scranton, Mississippi, by Delmas, who began grafting the tree into his nursery and orchard around 1890. When Delmas began shipping scions to D. L. Pierson in 1902, it appears that there was a mix-up, and several Delmas scions were taken for Schley, grafted onto the Summit Nurseries rootstock, and sold as Schley trees for several years before the problem was discovered. This was a sad discovery for many pecan growers in north Florida and south Georgia who, expecting the quality pecan found in a Schley, got instead a large nut with a thick shell, poor kernel quality, and poor disease resistance.

The nut production of Schley even as early as 1911 was considered variable, as reported in the *National Nurseryman*. Of course, this determination may be partially due to the mix-up with Delmas; however, in Georgia, under current management practices, Schley seems to be more productive in orchards within those counties bordering the Flint River than in those farther east. Many Schley trees were removed from orchards midway through the twentieth century because of the tree's susceptibility to pecan scab, which, without effective control measures, proved to be too excessive in this variety for commercial production. The same held true for the tree's susceptibility to a pest called the black pecan aphid, whose saliva contains a toxin that causes the development of yellow spots on the leaves that soon turn into brown necrotic patches covering the leaf surface. Without effective control measures, black aphids often result in defoliation of the tree. After effective fungicides, insecticides, and improved air-blast sprayers were developed and put into use by pecan growers in the 1970s, Schley once again became profitable. Although younger plantings of Schley can be found scattered here and there around the state of Georgia, they are very few in number and the variety is rarely planted today. However, many growers who have mature Schley trees growing in their orchards keep them in production and still obtain premium prices as a result of their exceptional kernel quality.

Schley is today often used in pecan breeding programs where pecan scientists seek its thin shell and high percentage of kernel. Schley can be found somewhere in the background of many pecan varieties in use today—and in up to 70 percent of the pecan varieties released by the USDA.

Although the Stuart and Schley varieties remain an important part of southeastern pecan production, another variety with roots in Jackson County, Mississippi, is today the pecan of choice in the southeastern United States. Charles Augustus Forkert was a Prussian born in Germany in 1854. As a young man, he had a great urge to wander and see the world. After four years as an apprentice to a florist and ornamental horticulturist, Forkert served three years as a soldier. After being discharged with honors, he traveled over Germany, Austria, Russia, and the United States from Massachusetts to Texas. The year 1883 found him living in Rochester, New York. A year later he made his way south, landing in New Orleans as a gardener in the Horticultural Hall of the Cotton Centennial Exposition.[31]

Forkert's horticultural skills served him well upon his arrival to the South. He eventually migrated to Ocean Springs around the turn of the century, when interest in pecans was beginning to boom. Forkert was in his late 40s before he began experimenting with pecan propagation. In addition to pecans, his early nursery efforts resulted in the sale of persimmon trees and grape vines. In 1910, Forkert purchased 53 acres of land in Ocean Springs, where he established Bay View Nursery along with pecan and fruit orchards. Charles Forkert came to be held in high esteem as an authority on the cultivation of pecans, grapes, peaches, and persimmons. He had particular success with the Georgia Belle peach and the Minnie, Ellen Scott, and Rolando grape varieties.

Like E. E. Risien, Charles Forkert was not content simply to gather and plant nuts in the hopes that a few would produce trees that made quality pecans. These two men took the process of cultivar development one step further. As Forkert put it in a paper presented at the 1914 National Nut Growers Association Conference in Thomasville, Georgia, "I concluded to try and assist nature in improving on some of them [pecan varieties] and ameliorate the good with the faulty ones."

Forkert describes his first attempt at cross-pollination: "It was in 1903, with the approach of the blooming season for the pecans that I concluded to try cross-breeding pecans. . . . Nine long years passed before trees grown from my first hybridized pecans began bearing. There were twenty three trees grown from the first lot. Two trees of this lot began bearing the ninth year, one a cross between Jewett and Pabst, which finally ripened some small inferior nuts. . . . The other tree, a cross between Jewett and Success, showed up much better. This tree bears a large nut, thin hull, and thin shell . . . cracking quality very good, kernel bright-straw-colored, plump, and smooth. . . . The nuts are heavy, going forty to forty two to the pound. The tree is very prolific and a healthy grower." [32]

Forkert appears here to be describing a variety he later named Desirable, and it was at this meeting in Thomasville, Georgia, that he may have first presented this important nut to other pecan growers. At the conclusion of his presentation, he produced a handful of nuts from his pocket and passed them around to those in attendance for inspection.

Another of Forkert's varieties, Dependable, was also, according to Forkert, a Success × Jewett cross. The two are very similar in appearance, which led to speculation that these varieties were siblings or were from the same cross. Yet another variety produced by Forkert at about the same time was Admirable, a Russell × Success cross. Genetic analysis has shown that Desirable appears to be of the same parentage as Admirable. Interestingly, following the death of Charles Forkert in 1928, his stepson-in law, Alceide Veillon, allegedly destroyed his notes rather than provide them to government researchers who were interested in his work on pecans.

Desirable apparently failed to make much of an impression on the crowd in Thomasville that day in 1914. With little early interest from growers, it was nearly lost to the pecan world. Despite the destruction of Forkert's records, the USDA managed to acquire scions from the tree in 1925. It began testing Desirable as US-7191 at the US Pecan Field Station in Philema, Georgia, near Albany. Were it not for this fortunate turn, pecan growers, particularly in Georgia, would never have had what has become the most profitable variety currently grown in the Southeast, and pecan production as a whole would not be the industry that it is today.

It was from the Philema USDA station that Desirable was disseminated, beginning about 1930, for trial planting in commercial orchards. Mr. R. M. Marbury Sr. was likely the first grower to establish a Desirable tree from the USDA station when he top-worked a tree in his yard in Albany, Georgia, to Desirable.[33] He later top-worked a large portion of his orchard to the variety. Desirable was finally introduced commercially in 1945, almost half a century after Charles Forkert's introduction in Thomasville. By the 1960s, Desirable was widely planted and continues to grow in popularity, despite a few challenges. When planting of Desirable began in earnest, it rarely developed pecan scab. Over the years, as the fungus has adapted to Desirable's defenses, the tree has developed a high susceptibility to the disease and can now be grown only under intensive management. Still, today it is the most commonly planted pecan cultivar in the state of Georgia, making up nearly 30 percent of all pecan trees planted in the state during 2012.[34]

What still makes Desirable so popular in the southeastern United States, even in the face of its cultural management challenges, is its consistent and nearly annual crops of large, excellent-quality pecans that shell out very well. High-quality Desirable pecans produce a nut with 52 to 55 percent kernel and are in the range of 40 to 42 nuts per pound, as suggested many years ago by Charles Forkert. For this reason, they sell for prices much higher than those obtained for Stuart. Still, Desirable is not without its faults. It remains very susceptible to pecan scab and cannot be produced consistently without regular fungicide applications during the growing season to control the disease. Growers must have the willingness, equipment, and pockets deep enough to apply the costly chemicals. For those who can grow it, Desirable is currently considered by many to be the most marketable pecan variety available.

Theodore "Theo" Bechtel of Ocean Springs was another prominent pecan nurseryman at the turn of the twentieth century. Bechtel was born to a family of German immigrants in Staunton, Illinois. While still in Illinois, Theo developed skills as an arborist-horticulturist. He eventually developed and owned a fruit orchard there where he raised apples and pears. Along with his brother, August, he developed the Bechtel's Double Flowering Crab Apple, which was introduced at the Chicago World's Fair in 1898.[35]

In 1899, the horticultural services of Theo Bechtel were sought by Dr. Homer L. Stewart, a native of Michigan, who had purchased a tract of land from Henry Colligan in 1891 with the idea of growing celery. Bechtel decided to leave his home and business in Edwardsville, Illinois, to come south in order to work for Dr. Stewart. After settling in Ocean Springs, Bechtel eventually sold his Illinois orchards to focus on the opportunities presented by his new life in the South.

Following Dr. Stewart's death in 1907 in Goldfield, Nevada, Bechtel was hired by the recently widowed Martha Holcomb to care for the 250 acres of pecan and fruit orchards she and her husband had purchased east of Ocean Springs. Apparently, even before his employment with Mrs. Holcomb, Bechtel had leased 30 acres from the widow in 1901 to establish a pecan nursery, and three years later he purchased the land himself. Becoming something of a surrogate son to the childless widow, Bechtel was legated Holcomb's home and other properties prior to her death.

Theo Bechtel was a man of many interests. Aside from pecans, he engaged in dairy farming as operator of the Success Dairy. At the time, milk in Ocean Springs was delivered to homes in pails. Always searching for improvements in any field with which he was associated, Bechtel began delivering his milk door to door in bottles, becoming the first to do so in the area. But, deciding that he preferred nurturing plants rather than cows, Bechtel sold the Success Dairy in 1914.[36] Despite many interests and endeavors, it was through his efforts with the pecan that Bechtel made a name for himself. Theo Bechtel quickly gained a reputation for his knowledge of plants, his Germanic work ethic, and his personal integrity, which made him a leader among the growing number of Ocean Springs commercial orchardists and nurserymen.

Among Theo Bechtel's many contributions to the fledgling pecan industry was the development of the Success pecan, a parent of Desirable. Found as yet another seedling tree growing in the same William Schmidt orchard in which Charles Pabst had found the original Pabst tree, Success attracted the attention of Bechtel, who propagated the tree in the spring of 1902. Observing the parent tree's prolific crops, Bechtel introduced and named the tree in 1903. Success had good nut size and would bear some of the heaviest crops observed anywhere. However, many of the advantages of this variety also gave rise to its disadvantages. The incredible productivity of Success led to poorly filled kernels in the heavy crop years, and once alternate bearing began, it was severe.[37]

Upon its introduction, Success came to be widely planted in the Southeast and was one of the region's most popular varieties. Bechtel's early advertisements called Success "the nut that fills at both ends." However, because of susceptibility to pecan scab, severe alternate bearing, and poor kernel quality, Success's popularity waned in the 1960s. Yet many old trees of this cultivar can be found growing in commercial orchards throughout the southeastern United States, particularly in eastern Georgia. Success continues to be used in breeding programs in order to harness its prolific cropping tendencies. In addition to Desirable, a very popular variety from the USDA breeding program, Pawnee, has Success in its parentage and is widely planted across the US pecan belt.

Bechtel's innovations in the area of pecan production were many. He introduced a number of other varieties besides Success, including Candy and Jackson. By adding a hardening rosin to his grafting wax, Bechtel provided a measure of protection for the grafts from the detriments of the humid southeastern climate. Theo Bechtel is also reported to have been the first pecan producer on the Mississippi Gulf Coast to use a chemical spray to combat pecan scab.

It is largely through the pioneering efforts of Antoine, Emil Bourgeois, E. E. Risien, and the Jackson County, Mississippi, nurserymen that the pecan has developed into a profitable, burgeoning, international crop. Today, although only a handful of the early cultivars are used extensively, there are over 1,000 recognized pecan varieties in existence. Many of the more recent releases (through the USDA and other pecan breeding programs after about 1950) are the result of controlled crosses and classical plant breeding. However, most of the early varieties from which the industry had its start arose from individual seedling trees whose qualities someone once noticed and set about propagating. The genetic diversity found in the many varieties of pecan today is the foundation of sustained and successful pecan production.

· 3 ·

The Secret Life of Pecan Trees

If one way be better than another, that, you may be sure, is nature's way.

—Aristotle

THROUGHOUT THE US COASTAL PLAIN from North Carolina into southern Texas, pecan trees inhabit the yards, roadsides, and orchards of the rural uplands and rolling hills. In the southwestern deserts of West Texas, New Mexico, and Arizona, pecans grow on tabletop flatlands shaped and leveled by lasers. But the pecan is a tree of the forest. It evolved through competition with other trees and plants, carving out a niche over time in which it could thrive. The pecan, like other plants, developed a set of characteristics suitable to life in its preferred environment, the bottomlands along streams from the Mississippi River Valley to east-central Texas, and southward into northeastern Mexico.

As intriguing as the history of pecan as a crop is, the story of this tree as a species is fascinating as well. The remarkable diversity found in the pecan holds the key to its success as a species. The varying environments throughout its home range have made the pecan tolerant of a wide range of conditions. The native range of pecan covers a wide geographical area extending from a latitude of about 42° north near Clinton, Iowa, southward to 16° north in Oaxaca, Mexico. At the northern end of this range the average low temperature in January is about 13° F, while the average high in the warmest month, July, is 85° F. Contrast this with the temperatures in southern Texas, where summer highs reach over 100° F and winter temperatures average a mild 40° F, and it is easy to recognize the ability of this tree to adapt.

Native pecan trees in different regions vary in their levels of cold tolerance. Those from the northern end of the tree's natural range have withstood temperatures as low as -40° F. For this reason, when pecan scientists and nursery workers seek a rootstock for use in northern growing regions, they generally seek seedlings from northern pecan sources. In this way, pecan growers can use the tree's natural adaptability to an advantage in growing pecans commercially. Lethal freeze damage to a pecan tree occurs most often near the soil surface, where air temperatures are usually lower. The protective effect offered by cold-tolerant rootstock can defend a tree from harsh winter conditions that it otherwise could not survive.

The main environmental factor limiting the extent of the pecan's native range is the length of the growing season, which is marked by the number of frost-free days. Although the major commercial production areas have a frost-free period of nearly 180 to 200 days or more, pecan trees at the northern end of their range can produce fruit within as little as 140 days. Pecan trees may still grow in areas with fewer frost-free days, but they will not produce fruit. This minimum time requirement for pecan fruit development appears to be under the control of the tree's genetic code. Although climate obviously sets an upper limit of time in which the tree's fruit can develop, the fruit development period is stable within individual trees sharing the same genetics and does not change with the length of the growing season.[1] For instance, consider for a moment that we find a pecan tree in Iowa that produces fruit within 140 days. If we remove graftwood from that tree and graft it into an orchard in central Texas, it will eventually still produce fruit within about 140 days even though it is exposed to a longer growing season.

Most fruit trees break dormancy and produce only under very strict chilling requirements, which means they require a certain number of hours below 45° F in order to break down growth inhibitors in their buds, which will then allow them to begin expansion in the spring. The number of chilling hours required to break dormancy varies considerably not only between species but within a species as well. For example, the number of chilling hours required for peach varieties may range from as few as 100 to as many as 1,200.

The pecan does not appear to be quite so restricted by chilling. As with most tree crops, the required number of chilling hours for budbreak within pecan varies by variety. Some studies have reported that 300 to 500 chilling hours were required for Desirable, Mahan, Success, and Schley budbreak, while Stuart required from 600 to over 1,000 hours.[2] Pecan budbreak will occur with fewer than even 100 chilling hours, but this may cause the timing of budbreak, even within the same variety, to be very uneven, leading to problems with pollination.

The chilling requirement of pecan is actually quite complex and not as cut-and-dried as it may at first appear. At the southern end of its range in Oaxaca, Mexico, the pecan may experience no freezing temperatures whatsoever in some years, and new growth often occurs before the previous season's foliage is shed.[3] It is believed that the actual chilling requirement for pecan can vary with fall conditions. If trees are exposed to cooler fall temperatures (less than 34° F), the intensity of the buds' rest or the number of chilling hours required for budbreak increases.[4] Therefore, in areas of mild falls and winters, pecans often experience a low rest intensity. As a result, trees of a single variety growing in areas with warm falls and winters require fewer chilling hours than the same variety growing in a colder climate.

It is interesting to note that buds on the ends of branches, called terminal buds, tend to have a lower chilling requirement than lateral buds, or those that run up and down the branch or shoot. Therefore, in years following cold winters, the tree will theoretically produce more shoots and potentially more nuts. It is actually the variation in budbreak that appears to be more sensitive to chilling than is the actual timing of budbreak. Thus, the main consequence of warm winters with low chilling hours is usually a sporadic, staggered, and nonuniform budbreak, which can generate problems with pollination and staggered nut maturity. The onset of new pecan growth in the spring is actually regulated by an interaction of both chilling and heating.

Of the two environmental factors, it appears that heat units in spring, rather than chilling units in winter, are more limiting to the growth of pecan. Under conditions of high chilling, minimal heat is required for spring budbreak, allowing the tree's shoots to begin growing as soon as possible. When growth does begin, it commences rapidly, with leaves and new shoots developing fully within about 30 days following budbreak. This increases the likelihood that the fruiting cycle will be completed within a relatively short period. In Hartford, Connecticut, pecans fail to produce mature nuts even though the length of the growing season is sufficient for fruit production. However, the low heat units accumulated there, particularly in early spring, serve as the primary obstacle to fruit production. In most cases, pecan production requires an average seasonal total of over 555 hours of temperatures above 65° F.[5]

A common observation in the southeastern United States is that the pecan is generally the last deciduous tree to break bud in the spring. This results from the relatively high heat unit requirement for pecan budbreak. As a result, people native to these areas often recognize the end of winter being signaled when pecan trees begin to leaf out. Some even use this as a gauge by which to determine that all danger of frost has passed and it is safe to plant their gardens and crops.

However, this rule of thumb does not always hold true. Pecans are sometimes damaged by late spring freezes, even in the southern regions of their range and growing areas. Still, the less sensitive response of pecan to low heating temperatures delays budbreak, minimizing the likelihood of damage from a late spring freeze. The pecan tree's sensitivity, or lack thereof, to heat and cold in relation to its budbreak and foliage development is a primary key to its ability to grow over such a wide range of geographic and climatic conditions.

In response to varying environmental conditions, the pecan has developed additional characteristics that allow it to survive across its wide distribution. As a result of the more mild temperatures, pecan trees from the southern extent of the native range tend to break bud earlier in the spring and hold their leaves later into the fall. For example, in Mexico, pecans may retain green foliage as late as January. Also, pecans originating from southern areas ripen their fruit much earlier in general than those from northern areas. Trees from these southern areas also have narrower leaf angles and wider limb angles, conditions that may allow them access to more sunlight than those with horizontal leaves in crowded situations such as wild forest stands. The leaves of pecan trees originating in northern areas tend to appear more reddish, particularly in Kansas and Missouri. Trees from northern areas have greater tolerance to freezing temperatures, even those as low as -40° F.[6]

One characteristic of pecan, common to most trees, is that as you move southward, the trees grow larger in both girth and height. This begs a few common questions: How big can a pecan tree get? How long do the trees live? How long do they produce nuts? How many pounds of pecans can a tree produce? The great height and massive girth to which some pecan trees grow can be remarkable. The largest pecan tree in the United States grew for many years along a rural roadside in Cocke County, Tennessee, near the town of Newport in the Great Smoky Mountains. Twin trunks emerged from a fork in the tree about 10 feet above the ground and spread their branches over a width of 111 feet. The tallest of these twin trunks rose to an official height of 141 feet. At chest height the unbranched portion of the lower trunk had a circumference of 267 inches, making it over 22 feet in girth. Sadly, the tree succumbed to a windstorm and lightning sometime between 2002 and 2008.

Currently, the National Tree Registry does not list a champion pecan tree. But a likely candidate is growing in Weatherford, Texas. The Weatherford tree, with a 19-foot circumference and a height of 118 feet, is estimated by the US Forest Service to be between 900 and 1,100 years old, although more likely, the tree is no more than 300 years old. Most pecan trees don't reach the age or size of these two champions. In general, pecan trees are relatively long-lived, capable of reaching ages of 150 to 200 years with heights of 80 to 100 feet in the absence of lightning strikes, windstorms, pests, or disease. Before its death, the record Tennessee tree continued to bear nuts into its old age, as does the Weatherford tree today.

The late Fred Brison, pioneering pecan horticulturist at Texas A&M University, remarked on an old "patriarch" pecan tree: "Other trees of comparable size in the same grove were cut down, and annual growth rings showed them to be over 200 years old. This tree produces 400 or 500 pounds of nuts for ordinary crops and one year it bore over 1000 lbs."[7] Such massive yields from a single pecan tree are not uncommon. In 1926, M. Hull of Louisiana State University reported a pecan tree 20 feet in circumference on the farm of G. B. Reuss at Hohen Solms, Louisiana, which produced as many as 1,600 pounds of nuts in a single season. Felix Herman of Bexar, Texas, swore before a notary in 1925 that he had gathered 2,200 pounds of pecans from an individual tree.[8]

One of the most crucial interactions of the pecan tree with its environment is that between the tree and water. Throughout much of the year, pecans teeter on a tenuous line between heat and soil moisture. A plant's temperature usually runs just above the air temperature. Plants don't sweat, but their major mechanism of cooling is through transpiration (water loss from leaves). Trees can dissipate tremendous amounts of heat when functioning normally. Hot temperatures often increase the dryness of the air, which causes the closing of small pores on the leaf surface called stomata. The stomata lead to a honeycomb of air spaces within the leaf that are filled with water vapor. When the stomata are open, water is transpired by the plant and is lost through the openings. When the stomata close because of hot temperatures, transpiration becomes limited and the plants can essentially overheat.

Features of the leaf surface such as the density of stomata and leaf hairs, termed trichomes, can have a strong influence on the tree's water-use efficiency. Pecan trees vary with regard to trichome and stomatal density. While trichome density may be influenced by environmental conditions and can vary even within a variety grown in different locations, stomatal density does not change simply based on where a tree is grown. This suggests that stomatal density is likely to be directly related to the environmental conditions of the location where that specific variety developed. Those pecan varieties with a higher density of stomata would likely be more adapted to hot, dry conditions than would those with fewer stomata.[9] These adaptations further represent the incredible diversity within the pecan that has allowed it to be grown over such a wide range.

Stomatal closure will not completely prevent water loss. Trees lose significant amounts of water directly through the leaf surface after the stomata close. Trees also lose water through lenticels (openings) on twigs, branches, roots, and stems. Heat stress within the tree depends on tissue temperature—the higher the temperature, the more water loss and the more heat stress the plant experiences.

A major effect of heat stress is the reduction of photosynthesis, that miraculous mechanism by which plants convert sunlight and water into sugars for energy. The reduction in photosynthesis brought on by heat stress is caused by a decline in leaf expansion, which in turn leads to less area for photosynthesis to occur, leaf loss, and an associated reduction in food production for the tree. When trees under drought are watered, photosynthesis may or may not return to normal. Recovery will depend on the tree's ability to adapt, the relative humidity, the soil moisture, and the duration of the drought.

The duration of hot temperatures for trees must not exceed the plant's ability to adjust, avoid, or repair problems. In this respect, pecans are much better adapted to handle heat than many other tree species. In fact, pecans require a heat unit accumulation of 4,500 to 5,000 degree days to complete nut development, and they grow best where the mean temperature is above 80° F from June through August.[10] Pecans also require warm nighttime temperatures for adequate nut growth. However, if high daytime temperatures create an unusually high demand for water, the tree may not have access to sufficient water to size the nuts properly, even under conditions of warm nighttime temperatures.

Pecans are relatively inefficient compared to other species when it comes to water use, requiring a large amount of water to support growth and fruit production. This, too, is a strategy developed by the pecan to lend it an advantage in its native environment. As a general rule, each 18° temperature step from 40° up to 130° F allows a doubling of respiration and water loss. Most plants become heat stressed at temperatures above 94° F, closing their stomata to conserve water. As a result, photosynthesis may double up to 94° F in most plants and then falls off rapidly. Pecans, however, under certain circumstances can tolerate higher temperatures of up to 106° F.[11] The key is soil moisture. As long as the tree has access to a sufficient water supply, its stomata remain open so that the leaves' gas exchange rates will remain normal, even under such high temperatures, allowing the leaves to function at normal capacity. This high tolerance for heat, which makes the pecan tree a very efficient transporter of water, is also the reason for its high water demand.

Under cultivation and with adequate irrigation, pecans grow quite well in the Desert Southwest. This is possible because of the pecan's incredible adaptability. The pecan tree's efficient water transport system evolved in native floodplain forests of the Mississippi River and Texas, where it had to compete with other species under conditions of high soil moisture and nutrient-rich soils. As a result, pecans have the ability to avoid stomatal closure under temperatures favorable for a high water deficit in competing species as long as there is adequate water available. So, as long as soil moisture is adequate, pecans can remain relatively cool when other trees slow down their photosynthetic rates.

Extremely high daytime temperatures, although undesirable, are not normally considered a serious limiting factor to pecan growth. However, under conditions of drought, less heat is required to damage the tree as the duration of high temperatures increases. Therefore, as the period of excessive temperatures is prolonged, the chance for some level of heat stress is increased.

When water is lost rapidly and leaf temperature increases, there can be a delay in water absorption by the roots. This can become a problem under conditions of poor soil moisture when leaves lose water faster than the roots can absorb it. Often a midday slowdown in transpiration occurs as the stomata close. Afternoon water shortages are corrected as quickly and efficiently as the soil water content allows via water uptake at night as tension in the water column pulls water into the tree. This is why soil moisture becomes so important. For some trees, nighttime uptake by roots can account for 20 to 40 percent of the plant's requirement.

Soil water limitations and heat buildup are closely linked and are especially critical in leaves. A common response to water stress is the shedding of leaves. Normal leaf shedding, as you see occurring in the fall, takes place via an organized leaf decline, which includes the loss of chlorophyll, causing the leaves to change color before they fall from the tree. With severe water stress, leaves may be shed while still full of valuable materials.

Sometimes leaf shedding may not occur until after rehydration. The cleaving of the leaf from the tree can be initiated by water stress but cannot be completed without adequate water to shear off connections between cell walls. This is why pecan trees may shed leaves or even fruit after a rain following a dry period. The oldest leaves are usually shed first. Injury to foliage and defoliation occur first in portions of the crown that are in full sun. These leaves may show drought-associated signs of rolling, folding, and curling. The actual physical process of knocking off leaves is associated with animals, wind, or rain.

Occasionally, plants or trees may suffer a sudden leaf death in which the leaves are killed before they can separate from the tree, causing dead leaves to stick to the tree. This condition often occurs when damage to the vascular system prevents the flow of water in the tree, most commonly as a result of cold damage. The tree's vascular system may be adequate to support it up to a point, but as temperatures rise, the tree overheats quickly.

In its native range, the pecan tree grows in areas that receive as little as 25 inches of rainfall to as much as 50 inches. In many of these areas, the rainfall received is much less than that required by the tree to thrive and produce nuts. The pecan tree circumvents this problem by growing along river bottomlands, which tend to have high water tables and a high potential for flooding. Yet, oddly, the pecan tree is sensitive to poorly drained soils, which lack the oxygen necessary for the tree's roots to function properly. So, how has this tree that is so intolerant of wet soils evolved to thrive in a river-bottom community? The secret lies in the tenuous three-way relationship between the pecan tree, water, and soil.

Pecans can tolerate water, and in fact they thrive on it, but only under certain conditions. Two important aspects of this relationship lie in when the trees are flooded and how long they remain so. Rainfall in the pecan's native range tends to be highest in winter and early spring. If flooding occurs in the winter, while the trees are dormant, they typically suffer no adverse effects. On the other hand, if flooding occurs when the tree is active, particularly at budbreak, root growth and subsequent shoot growth are reduced. Depending on the length of time the trees remain in flooded or waterlogged soil, the root depth may be so shallow that the trees are easily blown over by wind, or they simply decline as a result of poor root development and eventually die from lack of oxygen.[12]

The most vital link in this relationship between pecan trees and water is well-drained soil. Bottomland hardwood forests typically have distinct ecological zones at different elevations and flood frequencies. These zones are the result of an active, ever-changing river cutting its banks and forming new land. As one sets foot on land near a river or stream, the first land encountered is poorly drained and newly formed from soil and debris washed from upstream and deposited beside the river during flooding. As the erosion continues, more and more soil and debris increase the elevation of the land. Repeated flooding raises these "bars" above the normal high-water level, creating what is referred to as "front land." Subsequent floods overflow the front land and deposit coarse sediment near the river's edge, forming high, well-drained ridges.

Beyond the well-drained ridges the floodwaters settle, and between flood events these backwaters are inundated by tributaries. Fine sediment is deposited in these areas, and low, broad stretches of poorly drained soils form. The backwaters are often broken up by relief features such as low ridges, flats, and sloughs. Low ridges are formed by old, smaller streams and are generally more common than fronts or natural levees. It was here, on the rich, well-drained, loamy ridges, that the pecan tree found its niche. Although these ridges are subject to flooding, they do not typically remain flooded for extended periods. The coarser soil texture and slightly elevated topographical changes allow for good soil drainage that provides a zone in which the feeder roots of pecan trees can proliferate. The March to May rainfall and flooding pattern on these deep, well-drained soils coincides with the period of rapid growth flush, leaf expansion, seed germination, and seedling establishment. In many cases, the water table along these ridges is just high enough to allow the water-loving pecan tree a suitable source of water, but deep enough to allow good root development. In addition, the deep alluvial soils here are capable of retaining large volumes of water to sustain the tree for much of the season until the rains return.[13]

In some areas of the pecan tree's native range, the loamy soils surrounding the tree's roots may be several feet thick, a most critical aspect of the pecan's relationship with its environment. A seven-foot-deep soil has the potential to hold 9,000 gallons of water, enough to last a mature pecan tree for about 90 days, or from April through July. In this case, spring flooding would start the season off with a full soil profile that would be available to the tree until the late-season rains arrive. By contrast, a seven-inch soil can hold only about 800 gallons of water, enough for only 8 days. Thus, the pecan's complex and strict water and soil requirements confine it to a relatively restricted distribution within its river-bottom habitat.

Within the more arid regions of the pecan's native range, grasses, black willow, and eastern cottonwood are the major plants occurring alongside the pecan.[14] As one moves into more humid areas, hackberry, American elm, sycamore, green ash, sugarberry, and willow oak become more common in association with pecan trees. While pecans occupy the ridges, willows and cottonwoods occur on the newly formed front land. Sycamores are found higher up on the dry front lands, while the remaining species inhabit the poorly drained flats.[15]

Individual stands of naturally occurring pecan in native river bottoms may range in size from an acre or less to hundreds of acres. André Michaux, exploring along the Ohio River in the eighteenth century, reported a swamp of some 800 acres near the mouth of the Cumberland River where the trees grew in a nearly pure stand. Other reports exist of pecan forests of 300 to 400 acres along the Ohio River. The pecan tree is restricted to varying degrees by the declining soil drainage in the flats lying between the ridges. These flats are typically wide, with poor surface drainage. Mixed stands of trees occur here, with pecan growing mainly along the edges of the flats and near the ridges where the soil drainage is better, while other, more flood-tolerant species such as water hickory extend outward into the flats.[16]

Where the pecan occurs along the ridges of river bottomland, it is considered the "climax" species. This means that on these well-drained bottomland ridges, through the process of vegetation development over time (termed "ecological succession"), the pecan predominates in the final, self-perpetuating community of plants. In other words, the pecan is the species best adapted to such areas. The alluvial soils of the bottomland ridges abound with the richness of nutrients and life, a result of the abundance of organic matter in the form of forest debris and plant litter that is continually deposited there. Organic matter is nature's soil conditioner. This complex soil component has an overwhelming influence on almost all properties of the soil and can exist in several transitional states all at once.

The living portion of organic matter makes up about 15 percent of the total. It consists of a varied population of living organisms, including bacteria, fungi, algae, and plant roots. Indeed, the quantity of living material in soil is staggering. An acre furrow slice, or the upper 6.7 inches of soil in an acre of land, is said to contain between 5,000 and 20,000 pounds of living matter. The microorganisms in this fraction of organic matter help break down dead material, continuously recycling nutrients in the soil, which eventually become available for other plants to use. In forest soils such as the bottomland soils to which pecan has adapted, fungi dominate the microorganism community. Parts of these fungi, called mycorrhizae, produce a slimy material called glomalin that helps bind soil particles together, stabilizing them into clumps that constitute good soil structure, which in turn enhances water infiltration and prevents soil erosion. The mycorrhizal fungi develop a relationship with plants in which they infect the plant's root system and send out their own threadlike structures called hyphae that increase the plant's root contact with the soil, allowing the plant to acquire water and nutrients that would otherwise be unavailable. In return, the plant provides the fungi with sugars that are produced by the plant's leaves and transported to the root system.[17]

Fresh residues found in organic matter include recently deceased microorganisms, insects, fungi, old plant roots, sticks, leaves, and other debris that has yet to be broken down. In the pecan's native bottomlands as well as in commercial orchards, leaves, sticks, and shucks create abundant residue. It is this portion of organic matter that supplies food for the living portion. As fresh residues are decomposed, they release bound-up nutrients to the soil, nutrients that over time become available for plants. Most of the nutrients in organic matter are not in forms that plants can use. They must be converted from organic to mineral forms as they are decomposed by the life in the soil, a process called mineralization. In the river-bottom forests to which the pecan has adapted, this process takes the form of a full circle. The trees shed leaves, fruit, bark, branches, twigs, and roots. Sometimes entire trees are lost and decompose slowly over time. Most of this material remains on the bottomland floor. Occasional flooding may bring in debris from other areas upstream as well. As the nutrients in these living tissues are decomposed and mineralized in the soil over time, they are made available for use by the remaining trees and plants in the forest, and the cycle begins yet again. In this way, the cycling of nutrients helps enrich the earth and perpetuate this remarkable system.

Perhaps the most valuable component of organic matter from the standpoint of soil quality, if indeed one can be more valuable than the others, is the decomposed portion often referred to as "humus." It is here that vital, hidden biochemical reactions occur between plant roots and soil. This portion of the organic matter complex has been so finely decomposed by the activity of soil organisms that its individual particles are extremely small, providing a large surface area per unit of weight. In addition, the surface of this material is positively charged and attracts water and nutrients, which are bound there and prevented from leaching. A soil with good humus retains water for plant use and allows nutrients to be slowly released for uptake by plants.

The healthiest commercial pecan orchards attempt to mimic this aspect of the pecan tree's native environment. Much as they do in native pecan bottomlands, trees growing in orchards shed leaves, shucks, bark, sticks, and so forth. Aside from the crop itself, which is removed at harvest, and the sticks and limbs that are removed to make way for machinery to pass through the orchard, most organic debris that reaches the ground goes back into the soil, building organic matter and helping to fuel the system.

Although the magnified nut production in commercial orchards often requires additional fertilizers, there is a remarkable level of nutrient cycling in these systems as well. Only token amounts of nutrients are removed with the nut harvest. For example, a 1,000-pound-per-acre pecan crop removes only about 2.3 pounds of potassium and 1.6 pounds of phosphorus from the orchard in the form of shells and kernels.[18] This extensive cycling of nutrients often leads to difficulties for scientists attempting to study the fertilization needs of the pecan. In fact, nutrient turnover is so important in this perennial system that as many as six years of study may be required before scientists can detect any sort of decline in nut production or nutrient content in the tree itself where no fertilizer is applied.[19]

Most agricultural soils in the southeastern United States have an organic matter content ranging from 0.5 to 2 percent. The warm, humid climate of the region enhances the activity of soil microbes throughout much of the year, which prevents the accumulation of large amounts of organic matter, particularly in row crop fields where much of the soil is laid bare. Forest canopies, like those of the native bottomland forests to which pecan is native, help slow down the activity of soil microbes by keeping the ground shaded and cool. For the same reason, soil organic matter levels of pecan orchard soils often reach 3.5 percent or more, similar to the level seen in neighboring forest soils.[20]

As the tree is nourished by the soil, its roots perform another valuable service for the land itself. A pecan tree's root system is massive, extending sometimes twice the width of the tree canopy, often at considerable depths, anchoring the soil in place and protecting this precious resource from erosion. Additionally, many orchard soils in the humid growing regions from the Atlantic to East Texas are covered with grass or a combination of grass and clover. Thus these soils, particularly where clover is used, have another significant source of organic matter.

The addition of clover as a component of the agricultural system of a pecan orchard provides supplemental nitrogen for the crop and builds a healthy soil, providing benefits for the trees as well as for the environment as a whole. As the clover in orchards matures and produces seeds, it dries down and is mowed, returning the dried portions of the plant to the soil, enhancing the organic matter content of the orchard soil. This large amount of dry matter, rich in nitrogen, is ideally suited to the improvement of soils because it is composed of both carbon and relatively high levels of nitrogen. This allows the nitrogen to be made available for the surrounding plants, although it is released relatively slowly, which helps reduce its loss to the environment.

Prior to World War II, the planting of legumes like clover in the pecan orchards of the South was a common practice. Commercial fertilizers were unavailable or unaffordable for most farmers, so they sought inexpensive sources of nitrogen. The planting of legumes fulfilled that need. It is estimated that a solid stand of clover growing in an orchard can supply over 100 pounds of nitrogen, more than half that required by the tree itself. After the war, synthetic fertilizer was readily available and cheap. As a result, many pecan growers decided to forgo planting legumes and simply apply the cheap, reliable synthetic material instead. As fuel prices skyrocketed in the early twenty-first century, the practice of planting clover in orchards was renewed. Today, over half the commercial pecan producers in Georgia, the nation's top pecan-producing state, grow clover in their orchards.[21]

As a mature tree, the great height to which the pecan tree grows gives it a competitive advantage in the forested bottoms. This growth is directly related to the pecan's markedly long taproot. After a pecan seed germinates, the taproot quickly begins to grow and may reach a depth of six and a half feet at the end of its first season. During this first year, the aboveground shoot growth is minimal compared with that of the root. After about four years, when the root system has had time to become established, shoot growth explodes. The taproot continues to grow during this time, gaining access to soil moisture at depths that are unreachable to the pecan's shallow-rooted competitors. This allows the pecan to continue growing throughout the season when other species have stopped their growth because of insufficient soil moisture.[22]

Following a prolonged severe drought in the 1950s over the pecan's native range, it was discovered that pecan tree survival was enhanced under drought conditions for those with deeper taproots, which were able to reach down deep into the soil to extract the water needed to survive. The tree's taproot will grow to the depth allowed by good soil conditions. In most cases the taproot's growth stops when it reaches the soil's water table. As a result, some taproots can reach 20 feet or more in depth, maximizing the tree's anchorage in the ground. Such trees often break during high winds rather than uproot, allowing them to resprout and quickly continue growth. The remarkable ability of pecan roots to adapt to the level of the water table allows the tree to occupy large areas of deep soil volume, which in turn enables it to establish similar stand densities on semiarid and humid sites. In drier regions, wild pecan trees are restricted to narrow bands along streams where the water table remains high. At times the water table may be less than three feet deep. As the river level and associated water table drop gradually, the roots can grow downward and maintain contact with this precious resource. In more humid areas, native pecan forests may occupy broad sweeps of land, reaching into deep loamy soils. When extended droughts occur in these humid areas, conditions can change rapidly and the water table may drop below where the tree's roots have developed during better conditions. As a result, large stands of pecan may die.[23]

Droughts severe enough to cause the death of pecan trees are rare because the pecan tree is somewhat resistant to drought in spite of its high water requirement. The tree often responds to a severe drought year by developing short shoots and small leaves. This response reduces the tree's water requirement because it has less leaf area. If the drought worsens, branches begin to die in the top of the tree, progressing downward. Usually, the tree recovers relatively quickly when the drought ends, but when severe drought occurs in a year of heavy fruit set, the pecan tree is doubly stressed and may require as many as three years to bounce back. The years 2012 to 2013 saw a large number of pecan trees in Texas succumb to drought.

Lateral roots begin their growth in the second year following germination of the seed. These roots develop along the full length of the taproot, but it is the most shallow of these lateral roots, particularly those near the soil surface, that are essential to the tree's anchorage in the soil. The lateral roots of pecan trees can reach remarkable lengths of up to twice the canopy width of the tree or more in their quest for sufficient soil volume to meet the moisture demand of the tree's canopy.

From the lateral roots, vertical roots develop, growing toward the soil surface, where they begin to fan out in the soil's humus layer. In scientific circles, such roots are referred to as "humus strivers," but they are more commonly called "feeder roots." Feeder roots, as they are found in pecan trees, are absent from other deciduous fruit trees. These roots are small, usually less than one millimeter in diameter, and under ideal soil conditions, they form a dense, complex network of thin mats that scavenge the soil for nutrients and water. Feeder roots are usually developed by the fourth year following seedling germination. It is at this time that shoot growth and proliferation in a young pecan tree really begin to take off as the aboveground portion of the tree is fed by the extensive feeder root system.

Contrary to the roots of most plants, pecan roots do not have root hairs,[24] which give the roots more surface area and allow them to obtain more nutrients. In pecan, this function is carried out by the mycorrhizal fungi relationship described earlier. There are two basic types of these fungi, those that penetrate the root cells of plants, called arbuscular mycorrhizae, and those that grow on the outside of the roots, called ectomycorrhizae. It is this last group, the ectomycorrhizae, that colonize pecan tree roots. The spores of the fungi occur in the soil and infect the pecan feeder roots when stimulated by chemical secretions from the roots. As the fungus grows, it encloses the root, forming a dense sheath around it called a mantle. In addition to obtaining water and nutrients for the tree, the mycorrhizae form a barrier to protect the roots from harmful microorganisms and produce antibiotics to prevent infection.

Mycorrhizae help the tree roots acquire hundreds and even thousands of times more nitrogen, phosphorus, zinc, copper, and other nutrients than would be possible without them. The roots of pecan trees can actually be inoculated with the fungi when the tree is planted to help ensure fungal colonization. Pecan trees inoculated with mycorrhizae have been shown to grow faster and have higher levels of nutrients, such as nitrogen, phosphorus, potassium, calcium, and copper.[25]

A 2011 study detected 44 distinct ectomycorrhizal species in Georgia pecan orchards. One of the most commonly observed fungi associated with pecan trees in the eastern United States is *Scleroderma bovista*, otherwise known as the potato earthball because of its resemblance to a small spud. These fungi are inedible to humans. But one mycorrhizal fungus associated with pecan, *Tuber lyonii*, is a truffle related to the highly prized white and black truffles used as culinary treats in Europe. As an intriguing side note, two species of European truffle, *Tuber aestivum* and *Tuber borchii*, have recently been shown to produce well-formed ectomychorrizae on pecan seedlings.[26]

Mycorrhizae were described as early as the 1870s when H. Bruchman, a German biologist, determined that growths observed on the roots of pine trees by Theodor Hartig in 1840 were actually fungi. Still, most scientists of that time perceived fungi only as the main cause of disease and decay in plants, and the term "mycorrhizae" did not yet exist. However, in 1885, spurred by a commission from the king of Prussia to determine a method of growing European truffles, German botanist Albert Bernhard Frank interpreted the relationship between the roots of trees and certain fungi as mutually beneficial to both organisms. Frank had previously coined the term "symbiotism" (later changed to "symbiosis") in reference to lichens, which are a mutually beneficial relationship between fungi and algae. Regarding mycorrhizae, Frank wrote, "Certain tree species do not nourish themselves independently in the soil, but regularly establish a symbiosis with fungal mycelium over the entire root system. This mycelium . . . performs the entire nourishment of the tree from the soil. . . . The intimate, reciprocal dependence that follows the growth of both partners and the tight interrelationships of physiological functions that must exist between the two appear to be a new example of symbiosis in the plant kingdom." Frank deemed this relationship "mykorhizae," a combination of Greek words meaning literally "fungus-root." Although Frank failed to develop a practical method of truffle cultivation, his work brought to light a valuable relationship that exists in the plant kingdom. As Michael F. Allen suggests in *The Ecology of Mycorrhizae*, "We all should fail so nobly."[27]

Tuber lyonii, commonly called the pecan truffle, was first associated with the pecan in the late 1950s when C. Heimsch identified the fungus from a pecan orchard near Austin, Texas.[28] Since then, it has been found among pecan roots in Florida, Georgia, and New Mexico as well. The truffle has also been found intermingled with the roots of oaks from Mexico to Canada. Originally named *Tuber texense*, the pecan truffle has the appearance of a small, brown, lumpy potato, usually about the size of a golf ball.

Although they do not command the $600 to $1,000-per-pound prices of European truffles, pecan truffles have been known to fetch prices of $100 to $300 per pound from high-end, specialty restaurants in certain cities. They are usually found in well-irrigated commercial pecan orchards, particularly after drought years in which they are isolated in the more moisture-rich irrigated zones. The edible portion of the fungus is found unattached to the tree roots in the top inch or two of soil, anywhere from the trunk to the ends of the tree branches.[29]

If you're in the right orchard and you know what to look for, finding a pecan truffle can be about as hard as finding your favorite brand of cereal in the grocery aisle. Occasionally, the dirty tip of a truffle peeks through the soil. When one is found, there are usually more nearby. The high level of organic matter in pecan forests and orchards actually enhances the environment for mycorrhizal fungi, making conditions more favorable for this relationship. In the orchard system, this is particularly true where clover is used as a ground cover.

Dr. Tim Brenneman, a plant pathologist at the University of Georgia, has recently teamed up with dog trainers at Wynfield Plantation in Albany, Georgia, who are training Labrador retrievers to locate truffles in pecan orchards by smell. The Labs sniff their way up and down the tree rows, lying down when they come across one of the pungent mushrooms. But not everyone can afford their own truffle dog. Most of these dogs are taken from the same stock used to train bomb- and drug-sniffing dogs for the military and can fetch from $5,000 to $7,000.

As spring's new pecan shoots begin their growth aboveground, below the soil line the tree's roots are also beginning their annual growth and development. Root growth occurs first in the upper humus layers as soil temperatures begin to warm, reaching 65° to 70°F. As optimum soil temperatures reach down deeper into the soil and the saturated ground begins to drain, root growth commences at greater depths. At the same time, the soil in the upper layers is drying, bringing a temporary stop to the extension of corresponding roots.

When moisture in the upper soil layers is restored throughout the season, root growth begins again within 48 hours. Thus, the feeder roots constantly die and are replaced throughout the growing season as soil conditions allow. As the fruit of the pecan tree begins to mature, root growth at all depths is reduced so that the tree can focus its energies on the development of nuts.[30]

There are two particularly demanding periods in the annual cycle of a pecan tree's growth. In spring, as the initial growth of new shoots and foliage begins, pecan trees use stored energy from within to drive development of their new tissues. Once the new year's shoots reach approximately one-third of their final length (about three weeks after budbreak), there is a great demand for nutrients to replenish the energy lost to this exhausting process. Later, on the verge of autumn, pecans begin filling the developing kernels inside the shells of their nuts, another energy-exhausting process. In the pecan's native range, rainfall patterns correspond to these two distinct periods from March to May and again from August to September. Thus, the dense feeder roots of pecan trees in the shallow humus zone have access to good soil moisture in spring and late summer, allowing the trees to pull available nutrients from the shallow, nutrient-rich topsoil layer. These vital roots are most abundant at depths of 6 to 18 inches below the surface and are most prolific beneath the tips of branches, where drip from rainfall and leaf accumulation at shedding are likely to occur.[31]

The nutritional requirements of pecan trees are harmoniously linked to their feeder root system and its soil environment. While all plants require adequate supplies of the major plant nutrients nitrogen, phosphorus, and potassium, pecan trees have a unique affinity for certain micronutrients, which are generally required in comparatively small quantities by most living organisms to orchestrate a host of physiological processes. Although they are required in small amounts, the consequences of the absence of particular micronutrients can be severe. Pecan is highly sensitive to the unavailability of zinc, in particular. In its native habitat of soils rich in organic matter and humus, zinc deficiencies in pecan are rare. Only when these sites are disturbed by burning or when high temperatures accelerate the breakdown of organic matter (as a result of excessive tree removal) does zinc deficiency occur in native pecan bottomlands.[32] When pecan trees are not able to acquire adequate zinc, they weaken and produce telltale signs that something is not right. Among these symptoms is a condition known as rosette, in which the leaves turn yellow and take on a narrow, crinkled appearance. In addition, shoot growth is stunted and shoots and/or limbs may die. Pecan trees have a difficult time removing zinc from calcareous soils with high pH such as those found in Texas and the western growing regions. Along undisturbed native river bottoms, however, the rich humus topsoil provides a perfect environment in which pecan feeder roots can remove the precious nutrients the tree requires.

As it towers above its competitors, the pecan has access to sunlight and wind-borne pollen. Pecan trees bear both male and female flowers on the same tree, but these flowers generally do not mature at the same time, a trait bearing the cumbersome term "heterodichogamy." The flowering characteristics of pecan divide the many forms and varieties of the plant into two categories: those in which the pollen is shed before the female flowers are receptive (called Type I or protandrous), and those for which the reverse is true (called Type II or protogynous). This difference prevents self-fertilization, a botanical version of inbreeding that can lead to low vigor of the resulting offspring.

This characteristic of pecan is affected to some degree by climate. At the southern end of the range, where the climate is warm, maturity dates of male and female flowers on the same tree often overlap. In colder climates at the northern end of the pecan's range, there is virtually complete separation of the dates at which the male and female flowers mature. This complete separation ensures greater diversity in the population for valuable traits such as cold tolerance and early nut maturity that help the trees survive in harsh climates.[33]

As a pecan tree reaches sexual maturity, male flowers are borne on long, stringy tassels called catkins, which are produced a year or two before female flowers. This early dissemination of pollen holds no real advantage or disadvantage to the individual tree, but from the perspective of the tree stand or population, it ensures a ready pollen source once female flowers are produced.

In a sense, the pecan tree appears to value male flowers more than female flowers in that the production of male flowers is fueled by sugars or energy within the tree that is allocated from two years' worth of storage. By contrast, the development of female flowers is based on the energy from a single year. This difference serves two purposes. For one, the pecan population is assured of a better chance of a pollen supply even if adverse conditions occur. Also, in a stressful year, the trees can avoid the overproduction of flowers from which fruit could develop, potentially draining the tree of its energy stores.

Catkins occur at the base of new shoots and along the length of the supporting one-year-old wood. Pecan pollen is the bane of many an allergy sufferer, producing a preponderance of yellow dust that fills the air. Pecan, in fact, produces much more pollen than it actually needs. Each catkin stalk contains approximately 110 male flowers. As many as 2,000 pollen grains may be found on each individual flower. The height of pecan trees places the flower-bearing canopy at a level at which wind velocity increases the chances of pollination in mixed-forest stands. Pecan pollen may travel nearly 3,000 feet at a density sufficient to pollinate; however, in orchards and groves, the effective distance for adequate pollination between trees may be as little as 150 feet because of the density of trees between the pollinator and the target tree. Under cool, wet weather conditions, pollen shed is impaired. Once humidity drops, the male flowers release their pollen en masse, creating a dense yellow fog across native bottomlands and orchards wherever pecan is grown.[34]

Female or "pistillate" flowers are found on spikes at the tips of new shoots in clusters of 3 to 10. Unlike the showy flowers of peaches and apples, pecan flowers are rather inconspicuous, resembling tiny green starbursts, and blend in with the spring foliage so as to be almost unnoticeable. The stigma is that portion of the flower on which the pollen lands and germinates. The stigma is shaped like a cone and bears ornate extensions, which increase its surface area and enhance the efficiency with which it is able to collect pollen. There is a common misconception that when the stigmas of the female flowers of pecan turn brown, they are pollinated. This is not the case. Stigma color in pecan may be green, brown, pink, or burgundy, depending on the cultivar and age of the flower. Immature stigmas have a smooth, shiny surface. As they become receptive, they take on a moist, rough appearance. After pollination, the stigmas do turn brown, but a better indication of the end of pollination is that they dry up. Maximum nut set occurs when female flowers are pollinated within one day of becoming receptive. After four days, there is generally no fruit set. So, there is a window of about four days in which female flowers can be effectively pollinated and begin to produce nuts.[35]

Pollination is achieved by having the proper combinations of varieties within an appropriate distance from one another without significant obstacles in the way. Certain cultivars will serve as effective pollinators for others based on the compatibility between their pollen and female flowers. Determination of which cultivar pollinates which flower can be a tricky task. Historically, pecan scientists and producers have relied on charts of effective pollination dates for each variety to determine its appropriate pollinator. Most of these pollination charts are based on averages of pollen shed and pistil receptivity over multiple years. As a result, they contain a significant amount of variation. I have found that some years the charts will be right and some years they won't be, depending on chilling and heat unit accumulation in winter and spring, respectively. The annual variations in temperature during these seasons can lead to inaccurate pollinator choices.

Dr. Tommy Thompson, a retired USDA pecan breeder based in Texas, suggests that cool early-season temperatures result in a pronounced delay in pollen shed, but only a minor delay in pistil receptivity, so some patterns may not match in a year with cool early-season temperatures. An additional confounding factor is that as trees age, their pollination characteristics may alter somewhat. Flower maturity tends to be earlier in older trees, and the duration of pollen shed and pistil receptivity shortens as trees age. Also, the time interval between these windows shortens as trees age.

Another common misconception of pecan growers is, "There are enough seedling trees around to pollinate my trees." Every seedling pecan tree is unique. Just as 100 trees grown from nuts gathered from the same tree will likely produce nuts of varying size, shape, and quality, each one may also shed pollen at a different time. As a result, seedlings cannot be reliably counted on to provide adequate pollination in a commercial orchard setting unless you know when they shed pollen.

It is actually more important to consider how the pollinators are arranged in the orchard than it is to worry about how many pollinator trees you have. Pecan trees produce enormous amounts of pollen and need only a small fraction of what is produced within an orchard to ensure pollination, so it doesn't take many trees to do the job if they are in the right spots. Research indicates that as much as a 30 percent loss in fruit set may occur when trees are more than 150 feet away from a pollinator.[36] In most cases, proper pollination is more dependent on the amount of canopy between the pollinator and the target tree than on the distance between the two. In other words, as you get several rows away from the pollinator, the pollen is intercepted by other trees, causing the amount of pollen in the air to be diluted as it moves across the orchard.

There are two possible alternatives to effective cross-pollination: no pollination and self-pollination. Pecan undergoes four distinct periods of fruit drop in a given year. The first occurs early in the season when the flowers are at full bloom. These are often weak flowers that occur on short, weak shoots. The second drop occurs from 14 to 45 days after pollination. This drop is due to lack of fertilization of the egg. The third drop occurs at 54 to 90 days after pollination and is due to abortion of the embryo, often as a result of self-pollination. The fourth drop, occurring late in the season, is very subtle and has also been associated with self-pollination or incompatibility between the two parent cultivars.[37]

Estimates of self-pollination in pecan trees range from as little as 3 percent to as much as 49 percent, depending on the genetic diversity in the pecan stand, weather conditions, distance between trees, and so forth. Generally, when self-pollinated nuts do make it to maturity, they are of poorer quality than cross-pollinated nuts.[38] Germination of such pecans results in weak seedling trees, which, in their native forest setting, are less likely to endure the rigors of shading and the other stresses that can impact developing trees.

One of the more notable characteristics of pecan trees is that they are very slow to come into nut bearing. Although with optimal care, some cultivars may bear a few nuts in their third or fourth year in the field, six to eight years are required for pecan trees to produce a significant nut crop. This trait of the pecan, while something of a disadvantage for the pecan producer, lends the tree a competitive advantage in its native environment. Compared to its root growth, a pecan seedling's shoot growth is usually relatively minor during the first year. There is little branching during the first two to three years of growth, and the tree reaches almost straight upward in a phototropic race toward sunlight.

Beginning about the fourth year, branching occurs. This rapid upward growth and minimal branching reduces its early production potential but raises the tree to a position in the canopy at which it will have less competition for sunlight, one of its most valuable resources. The tree's functional geometry lies within a process called "apical dominance," which is the influence exerted by a terminal bud in suppressing the growth of the lateral buds below. This influence is controlled by a hormone called auxin, released by the uppermost or "apical" bud.

The biochemical mystery of auxin was unlocked in 1926 when plant physiologist Fritz Went isolated indoleacetic acid (IAA) from the growing tips of shoots. Went elegantly determined that IAA diffuses downward from the shoot tip, simultaneously elongating the shoot cells and inhibiting the buds below the apical bud. He also observed that the hormone moved from the illuminated to the dark side of the shoot, allowing the shoot to bend toward the light.[39]

The closer a lateral bud is to the apical bud, the more it is suppressed. If, by chance, the apical bud is lost through insect damage or wind breakage, for example, then the bud below will alter its course to become the tree's new leader. Trees "know" the value of a central leader. The tree's first priority for growth occurs at the tips or apices of stems. Lateral growth of branches is secondary to the tree's needs. In this way the tree reaches sunlight as quickly as possible. Such rapid growth also helps the tree develop a strong structure to support its potentially massive canopy as a mature tree.

Although a species of the forest, pecan is intolerant of shade, preferring instead open areas of full sunlight. As a hint to their natural proclivity for sunlight, where pecans now grow naturally outside their native bottomlands, particularly in the southeastern United States, they occupy forest edges or open fields rather than the interiors of bottomland forests. Compared to the pecan, most of the trees associated with pecan in its natural environment are vigorous growers. Thus, it would seem that the slow-growing pecan tree would be at a disadvantage in a forested environment where competition for sunlight is at a premium and faster-growing trees could shade out the pecan seedlings. However, the pecan tree's highly specialized preference for loamy ridges along the bottomlands allows it to avoid competition from most other bottomland trees, which prefer lower-lying clay sites. Pecan trees can survive in moderate shade but will reach optimum nut production only when exposed to full sunlight. Within native pecan forests, young pecan seedlings are only moderately shaded and exist in varying stages of growth. As trees fall or die and areas of sunlight are opened up, the fortunate seedlings awaiting an opportunity can quickly resume their arrested growth and occupy the newly opened space to become a bearing tree.

The canopies of most trees, including pecan, have a primary purpose: to collect solar energy. In order to use the sun's energy to its full potential and create their own energy in the form of sugars, plants have developed morphological, physiological, and biochemical modifications to enhance efficient photosynthesis. One such development is the contrasting characteristics of leaves growing in the shade versus those growing in the sun. Leaves on the outside of the canopy are completely exposed to the sun, while those beneath the outside canopy are shaded to varying degrees. Sun leaves are normally smaller than shade leaves because they don't need as much surface area to collect sunlight. The cell walls of sun leaves are also thicker and they have a greater vascular system for the transport of water and nutrients. One effect of all these differences is that leaves exposed to full sunlight have higher rates of photosynthesis than do those growing in the shade.[40] For pecan, shade leaves are only about half as effective as sun leaves in converting sunlight into plant energy. Interestingly, pecan leaves growing in the shade maintain their photosynthetic ability late into the growing season, when the photosynthetic capacity of sun leaves has been reduced by about 60 percent.[41]

In the wilds, just as humans have always favored them for their pleasing taste and nutritional qualities, pecans are highly favored by a host of forest animals, especially squirrels and birds. In fact, squirrels, crows, and blue jays all show a preference for pecans over oaks and other hickories. Although crow predation receives a lot of attention, and justifiably so, it is the blue jay that is considered the major avian dispersal agent of pecan. Blue jays prefer small nuts because of their relatively short beak compared to that of crows, which is about two and a half times longer. Jays have been known to carry three small nuts simultaneously, one in the beak and two in the throat. A single blue jay can consume or damage seven and a half pounds of pecans per month for each of the three months that the nuts are vulnerable to predation.[42]

Harvesting the nuts directly from the tree, blue jays may carry them three-tenths of a mile. Most animals that feed on pecans store them in "caches." For the blue jay, this is an instinctive reflex rather than a conscious effort to store food for lean times. Often, the sites chosen for caching by blue jays and other animals are poorly suited for pecan tree growth and establishment. Sometimes, particularly where birds are concerned, serendipity plays a role in natural pecan dispersal. The failure of blue jays and crows to transport nuts successfully to their cache can spread pecans to new areas if they are simply dropped in a suitable location. This can occur if the nuts are too large for the birds' beaks or if they have to fight off other birds in flight. Longtime University of Georgia pecan scientist Dr. Darrell Sparks has documented the establishment of pecan seedlings along bird flight lanes across a Georgia pasture adjacent to a pecan orchard.[43] Blue jays also require large perches when eating pecans because they must hold the nut with their feet while cracking it. As a result, nuts may be dropped accidentally along fencerows, forest edges, and hedgerows.

Along with blue jays, crows are among the most intelligent of all birds and are known to frustrate pecan producers by feeding heavily on pecans as the nuts approach maturity. As a result, crow shooting is a favorite pastime of many pecan farmers, who stand at the ready upon the most distant squawk. Blue jays eat mostly small nuts, but crows will eat either large or small nuts, with a seeming preference for the large ones. Large Desirable pecans with an oval cross section are preferred 3.5 to 1 over smaller, round Stuart pecans by crows. Crows are also known to carry pecans for up to two miles, as opposed to the shorter distances traveled by blue jays. Choosing densely vegetated, grassy locations, crows cache their food along fencerows and forest edges, where many are forgotten or escape consumption to sprout on their own.

While blue jays may consume many pecans, it is the crow that is scorned by the pecan producer. Its feats of pecan thievery and consumption are legendary. An individual crow may cause the loss of as much as three and a half pounds of pecans in one week. Anecdotal accounts exist of 10,000 crows descending upon orchards in a single day, stripping individual trees of as many as 300 pounds of nuts.[44] Like other pecan consumers, they prefer thin-shelled nuts and are adept at locating trees with such nuts in orchards and forests. Pecan farmers insist that crows prefer the earliest-maturing and thinnest-shelled varieties over others.

Both the gray squirrel and its larger, more colorful cousin, the fox squirrel, also covet pecans. The eastern fox squirrel is, in fact, considered the major mammalian disperser of pecans, and reports have suggested that the fox squirrel accounts for 37 percent of all wildlife damage to pecans. A single squirrel can eat, damage, or remove up to three and a half pounds of pecans a week and may hoard as many as 25 pounds of pecans per year. The fact that squirrels so highly value pecans is not a surprise. These arboreal mammals evolved alongside hickory nuts. They are such efficient dispersers of pecans because they often bury many small stashes of nuts well away from the tree on which the nut was produced, a practice known as "scatter hoarding," as opposed to "larder hoarding," in which seeds are stored in large stashes. Relatively few of these nuts develop into trees because the squirrels are so adept at relocating the hoarded nuts. Some estimates suggest they can indeed recover 99 percent of the nuts they store.[45] Squirrels often collect nuts before they become mature, chewing entire clusters free from the tree and dragging them into the woods. Culprits caught in the orchard are conspicuous by the green stains around their mouths from gnawing into the shucks. Hoarding results in loss and much frustration for farmers when the nuts are taken from their productive trees.

Over the course of its evolution, the pecan tree developed defenses when its offspring were consumed by birds, rodents, and insects, ravaged by disease, or outcompeted by other plants. In the continuously escalating arms race of evolution, the pecan tree has adapted strategies that help ensure its perpetuation. The pecan does this in a variety of ways with the patience that can be found only in a long-lived tree. In nature, pecan trees, like most organisms, do not live in a vacuum. Thus, they must interact with the world around them just as you and I do. The tree's health and survival become dependent on how it interacts, not only with animals that may consume the nuts it produces, but also with other plants as they struggle for the same space, sunlight, water, and nutrients.

Many organisms, including plants, produce chemicals that can "leak" out of their tissues in various ways to produce major changes in the survival, growth, reproduction, and behavior of surrounding organisms. This phenomenon is termed "allelopathy."[46] One such example can be found in penicillin, which is produced by a fungus that grows on organic substrates like seeds. The penicillin exuded by the fungus inhibits the growth of bacteria, which compete with the fungus for the same nutrients.

Many species of plants are known to chemically inhibit the growth of others around them. One of the most commonly recognized examples of allelopathy comes from the walnut, a close relative of the pecan, which produces a chemical substance called juglone in its leaves, hulls, and inner root bark. Other plants under or near a walnut tree tend to yellow, wilt, and die as a result of the juglone released into the soil from the walnut's roots and from decaying leaves and hulls that fall to the soil surface. The first to notice the poisonous effect of walnut on other plants was Roman naturalist Pliny the Elder in 77 AD. Juglone has been known to have a toxic effect on other plant species competing with the walnut tree.[47]

Close association with walnut trees while pollen is being shed in the spring can produce allergic symptoms in both horses and humans. Horses exposed to black walnuts may develop laminitis, an inflammation of the hoof. Affected horses become unwilling to move or have their feet picked up, are depressed, may exhibit limb edema, and have difficulty breathing. Black walnut husks have even been used to kill fish. Old-timers would gather the black walnut fruit into burlap sacks, tie up the open end, and put the sack out into a pond or creek. The chemical would leach from the husks of the walnut fruit, through the burlap sack, and cause the fish to die.

Juglone is also being studied for its potential in cancer prevention. It has been used to treat various skin conditions such as boils, cold sores, herpes, mouth sores, poison ivy, skin rashes, and so forth. Walnut bark has been used as a tooth enamel, and the juice of its husks as an expellant of worms and parasites.[48]

As a cousin to the walnut, like other hickories, the pecan produces juglone as well, although at lower levels than most hickory species. In the 1970s, juglone from pecan trees was shown to be toxic to the fungus causing its most serious disease, pecan scab. Understandably, those cultivars least affected by pecan scab had higher levels of juglone. Among cultivars tested, Stevens had extremely high juglone levels, while Frotscher and Desirable exhibited the lowest content.[49] The presence of juglone and other compounds within the pecan tree's tissues helps provide it with a certain level of defense against competing plants, diseases, and predators, aiding its survival and perpetuation.

Although juglone and other chemicals within the pecan are important tools in its survival and evolutionary development, the tree's primary method of avoiding overpredation has become one of its most well-known characteristics. Each autumn across the southern United States, vast acreages of pecan trees, wild and cultivated, may be filled with nuts, generating a pulse of resources for human and beast alike. More often than not, virtually all pecan trees over a large region, and sometimes throughout the country, have either large crops of nuts or small ones. This highly variable and incredibly synchronized reproduction in forest species is termed "masting." It occurs to some extent in oaks, hickories, and many fruit trees as well. Individual plants do not mast. Rather, masting is a group phenomenon that results when plants within a population synchronize their reproductive activity. The term "masting" is derived from the Old English term "maest," used to describe nuts of forest trees that accumulated on the ground.

At another level, among those concerned with pecan production, particularly when referring to an individual tree, the process is called alternate or, perhaps more appropriately, irregular bearing, and it is considered the most limiting factor in the profitability of growing pecans. Irregular bearing best describes the pecan fruiting process because alternate bearing implies a more predictable pattern of a heavy, or "on" crop followed by a light, or "off" crop. This pattern is not necessarily true for the pecan, as the regular cycle may be altered in some way by Mother Nature or management, leading orchards, groves, or individual trees to generate two or more consecutive crops.

Regardless of what it is called, evolutionarily speaking, large crops of nuts ensure that the trees produce more nuts than their predators can consume. Thus, there are plenty of nuts to survive and germinate during a bumper crop year. Additionally, the small crop years keep seed predator populations in check so that there are too few animals to eat all the seeds produced during the bumper years. Nut predators have influenced the evolution of the pecan tree in many ways. For instance, why are the nuts of most seedling pecans predominantly so small and thick shelled? Because predators (including humans) prefer larger, thin-shelled nuts that are easier to get to and provide more energy per unit consumed.

Native Americans, pecan farmers, and scientists have been aware of this boom-and-bust cycle in the pecan for a long time. But for years, everyone was puzzled as to the cause of the phenomenon. In fact, the biological process responsible for the regulation of irregular bearing generated a long debate among pecan scientists. Only recently have researchers begun to explain with more certainty how this confounding process is regulated.

Among scientists, there were basically two different camps regarding irregular bearing: the plant hormone theory or the carbohydrate reserve theory. The plant hormone theory suggests that chemical signals within the tree induce or inhibit flower production in pecan. The carbohydrate theory, on the other hand, suggests that stresses on the tree that reduce the store of its energy reserves will in turn limit fruit production. As it turns out, both theories work together to help explain the phenomenon of irregular bearing.[50]

Pecan trees determine how many female flowers they will produce in a given year about eight months before they can actually be seen on the tree. Sometime in late summer, usually in August, chemical stimuli are produced within the tree in the form of plant hormones or growth regulators that essentially stimulate the cellular development of the next year's vegetative or reproductive shoots. Pecan fruits are rich in substances similar to plant hormones called gibberellins, which are known to inhibit the following year's bloom of female flowers, often referred to as the "return bloom." It is the entire tree's crop load that affects the ability of the tree to produce a return bloom the following year.

However, leaves are believed to produce chemical stimuli themselves that support the development of the following year's female flowers. Pecan scientists have shown that the more leaves there are on the tree relative to fruit, the more the return bloom is enhanced. This may simply mean that there is a greater production of sugars within the tree to supply the energy needed for the following year's flower development. On the other hand, there could be a direct stimulation or stabilization of flower development from chemical signals produced within the leaves.

The chemical trigger stimulating flower development in pecan is thought to be vulnerable to events that may occur prior to its firing. For example, return bloom is highly sensitive to early loss of the tree's foliage, or premature defoliation. It was long assumed that the loss of energy in the form of carbohydrates or sugars in the leaves explained this relationship. However, the reduction in the amount of stored energy within the tree as a result of premature defoliation is actually quite small considering the great reduction in return bloom that occurs when the pecan tree loses its leaves too early. Therefore, most pecan scientists now believe that pecan leaves must produce some sort of floral promoter that triggers the development of the following year's female flowers.

Still, too much evidence exists for some sort of role played by carbohydrate or energy reserves in the tree with respect to the tree's flowering potential. The large taproot and extensive root system of the pecan tree have a great capacity for storing carbohydrates through the winter to be used as energy for the tree in the following spring. A lot of evidence suggests that an inadequate supply of these carbohydrates may have a direct negative effect on female flower production. These reserves tend to be depleted following a heavy crop of pecans on the tree. But the storage of these carbohydrates occurs after the chemical stimulation that occurs in late summer. So, how can they affect flower production? The answer may lie in the theory that carbohydrates do have an effect on flowering, but it is simply indirect rather than direct. For example, high carbohydrate reserves do not guarantee a large return bloom, but low reserves have been associated with a reduction in flowering. Recall that the flowering process in pecan is subject to change. As a result, if the tree does not have adequate energy to produce the flowers that it is "programmed" to produce, it may simply abort those flowers. Thus, irregular bearing in pecan is subject to regulation at two levels, both directly by plant hormones and indirectly by carbohydrate reserves.

Producers of apples and other tree fruit crops use chemical thinners to remove blooms and/or fruit from the tree in order to thin the crop. This helps them have more stable production from year to year. To date, no such methods have been developed for use on pecan. Managing the boom-and-bust cycle is a particularly difficult reality for pecan producers.

However, pecan scientists have developed a way to "trick" pecan trees into producing a crop of pecans every year. In the fall, trees begin dropping nuts as the shucks open and winds blow the nuts down. Unwilling to wait for this process to occur, early pecan producers began knocking the nuts from the trees with long bamboo poles. Following World War II, rudimentary equipment was developed using cables attached to a tractor to shake the nuts from walnut trees in California. This technology made its way into pecan production eventually and with many advances, large, self-propelled boom shakers were developed. Powered by a diesel engine and a tangle of hydraulic lines, the shaker has a long boom that can be extended out to reach individual limbs of the tree and a scissor-like head clamp that can grab hold of the limbs and shake. Such equipment allows pecan producers to shake nuts from the tree so they can be gathered at a suitable time to reach the desired market.

Based on the plant hormone theory, horticulture professor Mike Smith at Oklahoma State University developed another use for the pecan shaker in the early 1990s. Smith used the tree shaker to shake some of the nuts from the tree in midsummer, a horrifying sight to pecan producers at the time. Smith's idea was to remove some of the fruit at just the right time to help promote a return crop on the tree the following year, but leave enough on the tree for the crop to remain profitable in the current year. Because nut quality tends to be better in crop-thinned trees, there is very little loss of income in the year crop thinning is done. The following year, fruit-thinned trees come back with a good crop of pecans when they most likely would have had very little crop otherwise.[51] Although it has been a tough sell to pecan producers who for years have struggled to keep as many nuts as possible on the tree, Smith's idea worked, and it is slowly catching on, along with improved irrigation, fertilization, sunlight management, and varieties, as a means of stabilizing pecan production.

But how does the process of irregular bearing become synchronized over a large area to produce masting? As it turns out, no one knows for certain; however, there are several possible explanations for the synchronization of crop production among trees. One popular idea, which is similar to the carbohydrate theory mentioned earlier, is called the "resource tracking hypothesis." This theory requires that the number of nuts produced varies in response to the amount of resources available. For example, warm years with adequate soil moisture would potentially enhance photosynthesis and increase the availability of soil nutrients, allowing the trees more resources to produce a crop. On the other hand, cold, dry years could have the opposite effect. Since rainfall and temperatures are similar over large areas, at least regionally, this would seem to easily explain crop synchronization. The source of crop synchronization in pecan is poorly studied and needs further investigation.

There is some relationship between rainfall in one year and crop production the following year, but it may not be what you would expect. The University of Georgia's Dr. Darrell Sparks developed a method of predicting pecan production in Georgia. The previous year's rain for May, June, and July was the most important climatic factor influencing pecan production.[52] He determined that basically, the less rain that falls from May through July in one year, the better the nut production the following year. In fact, the current year's soil moisture has less effect on nut production because pecan production is dominated by the number of nuts per tree and not by nut size. The number of nuts on a tree is determined in the previous year, not in the current year. Nut size is influenced largely by soil moisture during the nut-sizing period, which occurs during the middle of the growing season. Pecans hate "wet feet," which tends to suffocate roots. Excessive soil moisture leads to a prolonged reduction in photosynthesis and suppresses leaf growth, which in turn suppresses flower formation the following year.

This still doesn't explain the whole story. However, research on the causes of masting in other nut trees may shed a little more light on the process as it occurs in pecan. For most trees, the evidence does not support the simple explanation provided by resource tracking. The size of most nut crops varies much more than the relatively small changes in temperature and rainfall from one year to the next. As with pecans, scientists studying oak trees in Missouri have observed that a good crop of acorns one year is usually followed by a small crop the next, suggesting some sort of mechanism, either the resource tracking hypothesis or the plant hormone hypothesis. Studies of oak tree growth rings have shown that trees tend to grow slowly in the year that a large acorn crop is produced and grow well the following year, when the acorn crop is small. Pecan tree growth rings in a given year, however, do not correlate well with crop load.

Some would still suggest that trees funnel large amounts of resources into flower and seed production during mast years and use those resources for growth in poor crop years. This would support an evolutionary response to some sort of environmental challenge. But if tree reproduction is under the control of environmental cues, how can it be manipulated experimentally as in pecan, simply by removing fruit? On the other hand, if pecan tree flower production is stimulated or inhibited by plant hormones in the tree, based on crop load from the previous year, how can so many trees over a large area become synchronized?

As mentioned earlier, the value of pollination for a tree is enormous, particularly for wind-pollinated trees like pecan. It doesn't make much sense from an evolutionary standpoint to produce huge quantities of pollen or flowers that could potentially go to waste. In most cases, a large crop of male pecan flowers is associated with at least a fair crop of female flowers. If different trees across a region produce a large crop of male and female flowers in the same year, they maximize their chances of pollination and minimize wasted pollen. Because wind-pollinated tree species generally produce such abundant pollen, it was long assumed that pollen was, for practical purposes, unlimited. The effect of distance and barriers between a pecan tree and its pollinator, as mentioned earlier, has been understood since the late 1990s. Around the same time, this phenomenon was discovered to be important for other wind-pollinated tree species as well and has important implications for explaining why trees of a single species synchronize their flower production.[53]

The theory of pollen coupling suggests that if the seed production of a tree is limited by the availability of pollen from another tree in a forest or in an orchard, different trees are coupled by the available pollen supply and would have a synchronized reproductive pattern without an environmental cue to trigger it. For example, a tree that flowers in a year in which only a few other trees flower would fail to produce many fruits because of a limited supply of pollen. As a result, it would not suffer an inhibition of flowering in the following year and its resources would not be heavily depleted. So it would continue to flower the following year and each year thereafter until the year that other trees in the forest also produced an abundance of flowers and pollen. In such a year, all the trees would experience a large fruit set, using up many of their resources. This effect would allow individual trees to synchronize their reproduction, despite the multitude of factors that could disrupt the physiological process that leads to flowering. Even without the effect of environmental conditions on the trees or crop, pollen coupling can explain the synchronization of reproduction in masting trees. The question that lingers here is, to what degree do individual trees depend on pollen produced by other trees a long distance away?

Environmental conditions are still important because factors such as temperature and soil moisture can directly affect flowering. The Moran effect is an ecological term that basically suggests that environmental fluctuations can account for the synchronization of various patterns between different populations. There is much truth in it, particularly regarding pecan trees, whether they are growing in an orchard or a forest. Late spring freezes can destroy male and female flower crops. Waterlogged soil can reduce photosynthetic capacity. Ice storms and tornadoes can ravage a tree's structure, destroying fruiting wood. Cool, wet conditions can limit pollination. All of these environmental factors can affect the production of fruit and could, in a sense, "reset the clock" for trees over relatively large areas. Still, crop synchronization over large areas cannot be explained simply by climatic factors alone, or by pollen coupling alone. It can be explained only if both pollen coupling and environmental influences are at work simultaneously.

The pecan tree's unique ecological adaptations have made this species remarkably suited to growth over an expansive geographic area. For the tree, the development of viable seeds for reproduction is the sole purpose for producing fruit. For growers, however, the production of fruit is the ultimate goal, and they strive for consistent production of high-quality nuts. Often, these two goals take diverging paths. However, scientists and farmers continue to adapt cultural management practices to meet the biological demands of this tree in orchard settings. The resulting ecological systems found in many pecan orchards are notably, and sometimes surprisingly, healthy.

Ernest Thompson Seton, early-twentieth-century American author and one of the pioneers of the Boy Scouts of America, once wrote, "Every tree like every man, must decide for itself—will it live in the alluring forest and struggle to the top where alone is sunlight or give up the fight and content itself with the shade." The pecan is a tree of the forest, yet literally and figuratively, it has not been content to remain there in the shade.

· 4 ·

The Rise of an Industry

A society grows great when old men plant trees whose shade
they know they shall never sit in.

—Greek proverb

INTERSTATE 35 RUNS SOUTHWARD from Duluth, Minnesota, to Laredo, Texas, dividing the country with its ribbon of asphalt. It is rare for a road, particularly something so impersonal as an interstate, to play such a prominent role in the story of a crop. But I-35 is in proximity to the 100th meridian. Although only an imaginary line splitting the United States in half, the 100th meridian is known as the edge of the Great American Desert, demarcating the humid East from the arid West. John Wesley Powell, leader of the first official expedition to explore the Grand Canyon, established this boundary in 1879. East of this line of longitude, annual precipitation is greater than 20 inches, providing the possibility of crop production without irrigation. West of the 100th meridian, it is all but impossible to grow crops without some sort of human-made water delivery system. The 100th meridian runs approximately 180 miles west of I-35 as the highway passes through Oklahoma and Texas, and for all practical purposes, the interstate serves as a modern, visible representation of the 100th meridian with regard to pecan production.

Whether by fate or design, as it reaches the regions of pecan production in the southern section of the nation, I-35 also serves as a line of demarcation for a change in the production practices employed by pecan producers based on the demands of the tree and its environment. Within the United States, most kinds of edible tree nuts are grown almost exclusively in their own respective states. Almonds, pistachios, and walnuts are produced primarily in California. Oregon grows filberts, while macadamia nuts are grown in Hawaii. The widespread, multistate pecan industry, on the other hand, faces a different set of challenges presented by the diverse landscapes in which the crop is grown. In most respects, the pecan industry is divided into two separate regions: the humid East and the dry West, which are defined largely by which side of I-35 the orchard may fall on. A case could be made for a separate northern production region north of Oklahoma. But the presence of soil moisture delineated by I-35 is a key factor in the particular approaches to pecan management. Pecan production in the East differs greatly from that in the West.

One of the most difficult things to relate about growing pecans is that, often, much of the management that is put into the orchard is not necessarily for the crop year in which it is done. The pecan, being a tree, is a perennial crop, and it has a good memory. You can't just get off to a bad start one year, plow the crop up, and start over. Everything the tree endures or benefits from affects the crop two or three years down the road. For this reason, although timing is extremely critical to growing anything, it is especially so for perennial crops like pecan. As the demands of a tree's physiology change throughout a year, management practices must account for these changes at the appropriate time. The regimen includes fertilizer application, insect and disease control decisions, irrigation scheduling, and crop load management. And management of each of these issues, along with many others, varies with the environment in which the crop is grown.

Most people who don't know any better think that the only requirement for growing orchard crops is to throw out a little fertilizer and come back at harvest time to gather the crop. I've even had friends tell me, "Oh, growing pecans . . . that's not really farming." In fact, the technical and artistic challenges of consistently producing an orchard crop, particularly one as finicky as the pecan, are quite complex. For consistent production, the trees must be sprayed for insects, disease, weeds, and nutrient disorders. The trees are fertilized, pruned, crop thinned, irrigated, and the orchard maintained for adequate sunlight. More importantly, all of this must be done at the appropriate time of year. The equipment required for growing pecans is very costly and requires, in most cases, at least 200 acres in cultivation to justify the expense.

Planting a pecan orchard requires vision. It also requires patience, money, and hard work. Texas writer John Graves gives an accurate description of a pecan planting: "Row upon row, the young pecans march out across the sandy loam, alternating now with ranks of fruit trees that help pay costs until the time when nut crops will reach commercial bulk. Pecans are a long-term investment, more for the profit of sons than of the fathers who plant them. In our razzle-dazzle, speculative economy, that kind of permanent planning for land stands out pleasantly."[1]

As the pecan became recognized by a growing nation, its wild population throughout the Mississippi Valley, Texas, and Oklahoma expanded eastward throughout the southeastern states. As the twentieth century dawned, wild or seedling pecans were scattered widely throughout the region as a result of human planting of seeds and seedling trees. Additionally, many trees were established through animal dispersal along fencerows and in the edges of local forests. Pecans became popular backyard trees in the South, providing shade and nuts to eat or sell. The growing farm population of the region began to recognize the value of the pecan for consumption on the farm and, occasionally, for profit. A typical south Georgia farm family in 1917 was reported to consume between 40 and 100 pounds of pecans per year.[2]

Many Southern towns planted pecan trees along the city streets. In the 1920s, as various charitable societies planned to develop a Road of Remembrance stretching from Michigan to Florida, the Georgia section was to be lined with pecan trees, with the intention of harvesting the nuts to pay for the road's maintenance; however, this hope never came to pass.

Even after improved varieties were introduced, many nurserymen failed to understand that pecans do not breed true from seed. This misconception led to common errors in planting by unknowing individuals such as Frank Owens of Coahoma County, Mississippi, who purchased 2,000 "improved" pecan trees for planting on his farm only to discover 14 years later that his orchard was made up of 2,000 mixed seedlings. A few of these actually produced good nuts, and one of these would give rise to the Owens variety, propagated extensively in Mississippi and Arkansas.[3]

As improved varieties were disseminated, there was a dramatic increase in the number of pecan trees across the South. Between 1910 and 1920 the number of pecan trees recorded in thirteen Southern states rose 70 percent from 2,951,217 to 4,748, 864.[4] The pecan had taken hold.

Today, in the humid Southeast, the capital of pecan production is the state of Georgia. Nicknamed the Peach State, Georgia is the nation's leading producer of pecans, pumping out an average of 95 million pounds of pecans annually on about 140,000 acres of bearing pecan trees. The Peach State now ranks third in national peach production, behind California and South Carolina.

Pecan production in Georgia, however, had a rather inauspicious beginning. A historical marker in Saint Marys, along the Atlantic Coast, reads: "FIRST PECAN TREES GROWN HERE ABOUT 1840—Grown from pecan nuts found floating at sea by Capt. Samuel F. Flood and planted by his wife, Rebecca Grovenstine, on block 47. The remainder of these nuts were planted by St. Joseph Sebastian Arnow in the north half of block 26. These first plantings produced large and heavy bearing trees, as did their nuts and shoots in turn. Taken from St. Mary's to distant points throughout southeastern states they became famous before the Texas pecan was generally known."

There are a number of questionable statements in this historical marker, not the least of which is that "they became famous before the Texas pecan was generally known." Perhaps this renown was true in Georgia, but the Texas pecan was well known before the trees mentioned here were producing nuts.

Although this historical marker would lead one to believe that Georgia's thriving pecan industry developed from a few nuts, the true story of how Georgia's pecan industry came to be is an even more intriguing tale. The period from the late 1800s through the early 1900s was a difficult one for Georgia as it sought to heal from the economic ravages of the Civil War. Amid its many problems was the need to diversify its agriculture. However, agriculture in Georgia, as in the rest of the South, was in the firm grip of a single crop.

The appetite of Georgia farmers for cotton following the Civil War bordered on lunacy. In 1885 Georgia's agriculture commissioner stated, "The average farmer is ready enough to adopt an improved variety of cotton or Indian corn; but slow to give place to a plant of a different species, or yielding a different product." Many refused to diversify their farms to anything but cotton despite a flooded world market and depressed prices. During the 1890s, the price of cotton dropped to an all-time low of four and a half cents per pound, yet the cotton planting continued. Most reports on the economics of growing cotton showed that it was being grown at a loss or at such a small profit that only the most meager living was obtained. This was due largely to the high cost of labor and fertilizer.[5]

Actually, the continued planting of cotton in the face of obvious economic failure was often more a product of the banking systems than the farmers' desire to grow cotton. Land was worth almost nothing at the time, and banks would not accept much of anything except a cotton crop for the security of a loan. Thus, in an era where little capital was available, most farmers were left with little choice but to grow cotton. Not only did cotton production drive many a farmer to poverty, it also mined the land itself into poverty by stripping it of vital topsoil and nutrients. In addition, land speculation was rampant, particularly in southern Georgia, a frontier that was still being populated with all kinds of settlers, including northern industrialists seeking new economic investments.

In truth, Georgia's pecan industry is due to low land prices resulting from the failing cotton economy, exaggerated reports of income from pecan orchards, and a plethora of unsuspecting northern investors around the turn of the twentieth century. The heart of the state's pecan industry lies in and around the city of Albany in southwestern Georgia. One report in 1924 suggested that "there are now more than 50,000 acres in pecans within a radius of 40 miles from Albany," and that each year one million new pecan trees were estimated to be planted within a 100-mile radius of the town.[6] One testimony to the pecan business during this boom period was the Albany District Pecan

Exchange, organized in 1918. Occupying a factory building and warehouse on the corner of Washington and North Streets, the exchange served as a cooperative marketing association. During its first year of operation, the exchange handled only 70,000 pounds of nuts. By 1921, this number had increased to 500,000. The Albany District Pecan Exchange operated until 1949.

The small town of Baconton in Mitchell County, just south of Albany, lays claim to being "The World's Pecan Center." It was here, in 1880, that the daughters of Major R. J. Bacon received a barrel of pecans from their aunt, Mrs. O. L. Battle. Major Bacon was a prominent figure in the affairs of the area and is sometimes called "the prophet of the pecan industry" because of his belief in the possibilities of the crop. In addition, Mr. Bacon foretold the shelling of nuts by machines and the extraction of oil from the nuts.

After Major Bacon's little girls had planted the nuts along a fence in the garden, the trees grew and began to produce nuts reported to be of superior quality. From the further propagation of these trees, George Meriwether Bacon, nephew of R. J. Bacon, obtained five trees that he planted in the mid to late 1880s on his Pine Bluff Plantation about four miles north of Baconton, near DeWitt. From this stock, Bacon planted and grafted trees, increasing his pecan acreage to 700 acres. In the process, this planting grew to become the first cultivated pecan orchard in Georgia. Mr. Bacon and others soon established pecan nurseries to supply those interested in planting pecan trees.[7]

In 1905, the Jackson Pecan Grove Company, owned by R. P., W. T., and T. S. Jackson, along with C. C. Batey, began planting pecan trees in the area. This operation was run primarily from a Chicago sales office. The Jackson Company's plan, along with the plans of several others at the time, was to plant large acreages of pecans, which were sold to northern investors in smaller units. Once this promotional plan began to take off, scores of companies popped up around the Albany area guaranteeing profits from pecan production. Most of these companies planted the orchards and managed the trees for the first five years, at the end of which the purchaser would take over all responsibilities for the orchard's ownership and management.[8]

The pamphlet of one such company suggests,

We offer you a five acre pecan grove, on easy monthly payments, in Dougherty County, Ga., closely adjacent to Albany, Ga. We plant for you two year old, selected standard varieties of paper shell pecan trees, the finest nut trees in the world. When set out these trees are from two to three feet in height.

 We cultivate, fertilize, prune, and care for your trees for five years, and we guarantee to deliver to you, with perfect title to land, a grove of twenty pecan trees per acre. . . . Twenty dollars monthly, a total sum of $1260, will purchase one of our five acre groves. Each five acre grove should in a few years yield an annual income above all expenses of $1000 a year, and the profit increases annually thereafter. . . . We guarantee the land to be the best in the world for pecan culture; and where pecan seedlings just naturally grow wild.

Another brochure suspiciously boasts, "A pecan grove of 5 acres nets $2500 yearly with no worry, no loss of crop, and little cost of upkeep. The papershell pecan begins bearing at 2 yrs of age, producing 50–250 lbs at 10 yrs, with yearly increases thereafter . . . 5 acres will keep the average family in comfort."

The result of these pecan land schemes was speculative fervor in which orchards were planted and quickly sold with the purpose of making money through selling orchards rather than through selling the crop. The unknowing purchasers took over the management of their orchards at about the time that management became difficult. As a young tree, a pecan is easily managed. As long as it is kept free from competition with weeds and receives water, a little fertilizer, and some pruning, it grows very well with minimum effort. Still, the years to a productive tree can be many and would certainly have been so in that time before irrigation. Once pecan trees reach maturity, they tend to become a little more fickle. Orchard owners soon discovered that the trees did not produce as advertised, nor did they produce every year, and few families were "kept in comfort" solely on the income from their five-acre orchard.

 Through this process, thousands of acres of worn-out cotton land around

FIGURE 3 Pamphlet from the Keystone Pecan Company used to promote investment in Georgia pecan orchards in the early twentieth century.

the Albany area were planted to pecans. As the Depression hit and the owners of all those little five-acre orchards suffered hard times, their tracts were lost to creditors and/or were sold at a pittance to those few who could afford them. In this way, large pecan acreages were amassed. Many of these orchards are still in production, and today pecans form the tree canopy of Albany's residential areas and the surrounding countryside.

One such area occurs just east of US Highway 19 between Albany and Baconton near an area known, rightfully so, as Pecan City. Here lies what is likely the largest contiguous block of pecan orchards in the state, encompassing over 5,000 acres. Separated only by roads, this seemingly endless orchard is actually composed of multiple orchards under the ownership of six or seven individuals or farming operations. Here, you could walk nearly five miles in one direction and never be out from under the shade of a pecan tree.

The heart of Pecan City lies along Honeysuckle Road between Highway 19 and Highway 133, where four of these large orchards meet to form a sort of "four corners." Today, the orchards of Pecan City are some of the most profitable pecan operations in Georgia. Each contains a mixture of a number of pecan varieties, although the modern southeastern pecan trifecta of Stuart, Schley, and Desirable still dominate. The success of the orchards around Pecan City lies in the foresight and willingness of the orchard managers to implement progressive management techniques like mechanical fruit thinning, tree hedging, and selective orchard thinning, the latter of which removes unproductive or unprofitable trees from the orchard to open up sunlight and provide more resources for the remaining trees, sort of like thinning a cow herd. Most of the orchards making up the Pecan City block were planted during the period of speculation that gave birth to the Georgia pecan industry; however, they are frequently renewed by the planting of young pecan trees to replace old downed trees or unproductive varieties.

At the same time orchards began to be planted in the Albany area, pecan trees were being planted throughout the remaining southern half of Georgia, mostly in smaller acreages. Many of these orchards were planted in those counties lying along the Flint River from Bainbridge to Fort Valley, with Albany in the center. This area's sandy loam topsoil and underlying clay subsoil are well

FIGURE 4 An early-twentieth-century pecan orchard planted in Georgia. The pecan industry in the nation's top-producing state had an inauspicious beginning as an experiment in land speculation.

suited to pecan growth and production. Pecans were planted farther east of the Flint River as well. Orchards sprang up around the towns of Statesboro, Fitzgerald, Waycross, and Valdosta, some of which were planted through the same type of money-making schemes as were seen in the Albany area.

The many unfounded claims of profitability combined with an inadequate knowledge of pecan management hastened the end of the Georgia pecan land schemes. The fervor soon cooled and interest in the crop declined. Still, Georgia's pecan production slowly grew from around 3 million pounds in 1919 to 30 million pounds at the time of World War II. By 1950, Georgia was leading the nation in pecan production and remains in that position today.

Much of the credit for the survival of Georgia's pecan industry through the failed period of the pecan land schemes lies with a handful of men who made an honest attempt to develop the pecan tree into a viable crop for Georgia and the southeastern United States. On November 21, 1901, R. J. Bacon, his nephew G. M. Bacon, J. M. Tift, and J. M. Wilson met in the office of R. H.

Warren on Broad Street in downtown Albany. By the time this meeting had ended, the Southern Nut Growers Association had been formed. With the aid of others, including B. W. Stone and J. B Wight, this organization grew to become the National Nut Growers Association. As a voice for the organization, George Meriwether Bacon founded the publication the *Nut Grower* in August 1902 to serve as a source of information on the pecan and its culture. Throughout the early twentieth century, these men and others interested in pecans throughout the South were becoming organized. The early reports from the National Nut Growers Association, which would eventually evolve into the currently active Southeastern Pecan Growers Association, illustrate how the individuals associated with these groups wisely led the development of pecan production as an industry. Many of these early pecan growers conducted their own experiments and in a sense became the first agricultural scientists working on pecan in the Southeast.

One of these men, J. B. Wight of Cairo, Georgia, published a 16-page bulletin in 1906 entitled *Pecans: The What, When, How of Growing Them*, outlining what he had learned from 20 years of growing 120 acres of pecans. A later pamphlet by Wight, entitled *The Pecan: Some Points, Pointers and Suggestions*, continued to champion pecan production in elaborate prose: "When the wheat and corn fields of the West cease to be profitable, when there is no longer a demand for the fleecy staple of the south, and when the spindles of our factories lie idle because there is no need for their products, then . . . need the pecan growers fear for their industry."[9] In this particular pamphlet, Wight shares the growth and production records from one of his Frotscher trees from its planting in 1892 to 1915, describing it as "the best known pecan tree in existence due to the fact that the carefully kept records of its growth and its crops of nuts have been widely published." Wight describes the tree as being "bought in January 1892 from William Nelson, New Orleans; cost $2.00; height when set three feet; height in 1915, sixty six feet; spread of branches, eighty five feet. After this tree came in to bearing, stable manure and guano . . . have been applied about the tree each winter. . . . This tree has yielded an average income of $100 a year for five consecutive years."[10] Not bad for the early twentieth century.

The avoidance of frauds in the pecan business was of major interest to the early pecan organizations. In fact, the first article published in the *Nut Grower*, entitled "Unscrupulous Dealers," focused on the topic of fake budded and grafted trees. Each of these organizations held annual meetings and published proceedings, beginning with the first National Nut Growers Convention held in Macon, Georgia, on October 6, 1902. G. M. Bacon presided over the meeting, which boasted exhibits of nuts, nursery stock, and farm implements. In addition to topics such as grafting and pecan culture, presentations were made on the avoidance of frauds and fakes.[11]

R. J. Redding, director of the Georgia Agricultural Experiment Station in Experiment, Georgia, near Griffin, reported at the time that the knowledge of pecan culture was "chaotic and unsettled." He felt that organizations such as the National Nut Growers Association could do "the greatest good [for] the greatest number" because much of the information regarding pecans to be found in newspapers across the country was of "doubtful authenticity." Redding suggested that frauds and fakes be kept out of the new association in order to provide interested parties with a legitimate source of pecan information.[12]

Another interesting bit of commentary to come out of the National Nut Growers Association in its first year of operation was a committee decision on the correct pronunciation of the word "pecan." As an insight into their Southern dialect, the committee voted that the letter *a* in "pecan" be pronounced as in "can" or "fan." In addition, they voted that a planting of pecan trees be referred to as a "grove" rather than an "orchard."

One rather curious practice employed in the planting of trees early in the twentieth century was the use of explosives. Trees tend to grow better when planted in loose soil that allows roots to become established unimpeded by compacted soil layers. The DuPont Company distributed a brochure in 1911 entitled *Farming with Dynamite: A Few Hints to Farmers*. A few of the suggested uses of dynamite around the farm included clearing land; breaking up hardpans; draining swamps; digging ditches, post holes, wells, and reservoirs; making and grading roads; excavating foundations and cellars; trenching; regenerating old, worn-out farms; and planting and cultivating orchards. One testimony by Mr. S. H. Bolinger of Shreveport, Louisiana, describes the use of dynamite in blasting the holes in which 1,080 pecan trees and 8,000 peach trees were planted.[13]

Bolinger explained, "A 2 inch hole was bored about 4 to 4 ½ feet deep. In the auger hole one stick of 40% dynamite with fuse attached was inserted, the hole filled and lightly tamped; then exploded. The explosion created a space of about the size of an ordinary barrel. The ground was not blown out but was simply raised on the top about 3 or 4 inches. In almost every case, however, it could be seen that the ground had been thoroughly loosened up for a distance of 10 to 15 feet around the hole."[14]

As odd and risky as the use of dynamite in planting orchards sounds, it was a common practice in the early 1900s and was embraced even by university experts. A. J. Farley of the New Jersey Experiment Station in New Brunswick began testing the claims of the dynamite manufacturers on apple, peach, and pear tree plantings. Farley's studies revealed that peach trees planted with dynamite exhibited greater growth and smaller circumference. By the end of the second season there was no difference in size with and without dynamite, but apple and pear trees planted with dynamite developed deeper and stronger root systems.[15] Later studies by Farley generated conflicting results. One study suggested more growth and greater yield on trees planted with dynamite than on those planted without it. Another study suggested greater growth for trees planted in nondynamited holes. Farley's final conclusions on the subject were that "in the majority of cases the increased growth and production recorded on dynamited trees is not great enough to make up for the increased cost and danger involved in planting. Furthermore the use of dynamite is not recommended for tree plantings on those soils that are naturally adapted to orcharding." This was probably a wise suggestion.

As R. J. Redding had pointed out, knowledge regarding proper management of pecan orchards was sorely lacking. Following the pecan planting boom of the early twentieth century, a flurry of studies took place to help make the pecan a more profitable crop. Probably the most widespread cultural management problem facing pecans was a malady called rosette, which caused leaves to be small, yellowed, and disfigured into twisted and curled shapes. Affected trees turned pale green or yellow and branches began to die, particularly in the upper tree canopy.[16] The problem remained absent on the pecan's native soils, but as orchards were planted on upland soil from Texas eastward, rosette ravaged pecan trees. As early as 1902, the US Department of Agriculture was receiving requests for investigations into the cause of rosette. For many years, the answer to this problem eluded investigators.[17]

Rosette was believed by many to be a disease. Although various fungi and bacteria were isolated from branches and leaves showing rosette, no conclusive evidence as to its cause could be found. Investigations were made into varieties, budding and grafting, soils, and potential insect associations with rosette but still no definitive cause was found. Various fertilizers were applied in many experiments but none showed significant results. The closest the researchers came to discovering a cause was to notice that an application of lime seemed to make the problem worse, and, occasionally, iron sulfate or copper sulfate applied to the soil around the tree improved its appearance. It was a true agricultural mystery.

As with so many discoveries in science, the answer to the cause and treatment of pecan rosette came quite by accident. Investigators were puzzled as to why iron sulfate seemed to sometimes improve the condition of trees with pecan rosette, yet at other times it had no effect. In June 1932, A. O. Alben, J. R. Cole, and R. D. Lewis, with the USDA Bureau of Chemistry and Soils and the Bureau of Plant Industry, reported on an experiment in Louisiana with iron sulfate in which they dipped pecan leaves in iron sulfate and iron chloride solutions.[18] As in other experiments, some of these leaves improved significantly while others did not. Later that year, this group of researchers discovered why.

Alben, Cole, and Lewis published another small, preliminary note in the scientific journal *Phytopathology* in December 1932. Here they reported that their previous results with iron sulfate were favorable only when galvanized iron containers had been used to handle the solutions in which they dipped the pecan leaves.[19] When iron containers are galvanized, they are electrically coated with zinc, which prevents the metal from rusting. As a result, the iron salts used in the experiment contained "appreciable quantities of zinc." Accordingly, the researchers conducted another experiment using shellacked crocks in the place of the galvanized containers. This time zinc sulfate and zinc chloride were compared with iron sulfate solutions. The results showed that the zinc solutions restored the leaves to a healthy condition while iron sulfate had no effect. This brief note in a scientific journal reported one of the most important findings related to the production of pecans. With this serendipitous discovery, the worst problem in pecan production was overcome.

At the same time that Alben, Cole, and Lewis were realizing the relationship between zinc and pecan rosette, experiments were underway in Arizona that would show the same relationship. A. H. Finch and A. F. Kinnison from time to time saw positive results with the application of iron sulfate. Remarkably, they too discovered that when iron sulfate cured rosette, it was contaminated with zinc and that leaves with rosette contained less zinc than those without the disorder.[20]

Zinc is required to obtain good vegetative growth, vigor, and nut production of pecan trees. Without Alben, Cole, and Lewis's discovery, along with that of Finch and Kinnison, pecan trees from Texas to the Atlantic Coast would have failed to be productive. Many would have lost vigor and would likely have been cut down or died, taking the fledgling pecan industry along with them. Zinc deficiency is rare in the pecan's native habitat. The soil in the pecan's native area has a relatively high pH, which tends to inhibit the uptake of zinc by plant roots. This led to the worsening effects of rosette with the addition of lime in many early experiments. Although the high pH of the pecan's native soil would seem to be limiting, the abundant organic matter found there is rich in nutrients, including zinc, which allow the tree to flourish under such conditions.[21] A flurry of investigations on the effects of zinc on pecan trees quickly followed the initial discovery. It didn't take long for word to spread that an application of zinc to pecan trees would cure pecan rosette.

Even after the discovery of zinc deficiency as the cause of rosette in the

1930s, pecan producers in the arid lands west of the Mississippi still had difficulty with the disorder. They knew the cause of rosette, but simply applying zinc to the soil around a pecan tree did not always solve the problem. Just as the addition of lime to orchard soils revealed, the difference is based solely on soil chemistry. In the humid southeastern United States, the acidity of the soils in the region allows heavy metals such as zinc to be taken up readily by plants, so soil applications of zinc can often cure rosette relatively quickly. As mentioned previously, the alkaline soils of the arid Southwest, however, often bind metals in the soil so that plants cannot remove them. This condition sometimes made it difficult to correct rosette with soil applications.

The answer to the problem came in spraying the leaves of the tree with solutions of zinc. The greatest champion of this practice was Texas A&M horticulturist Benton Storey, who, among his many contributions to the improvement of pecan production, worked extensively to find the best method of getting zinc into pecan trees. He determined that spraying the leaves as they were developing allowed the tree to absorb the zinc directly into the tender foliage where it was needed, eliminating the problem of rosette where pecans are grown on alkaline soils.[22] Foliar zinc sprays were found to be so efficient at curing the malady, they became a recommended practice wherever pecans were grown.

The rate of pecan production in the Southeast skyrocketed following the discovery of the cure for rosette. The dramatic effects of this discovery are easily observed in the historical production records from Georgia. In 1932, Georgia produced only 5 million pounds of pecans. Four years later, once producers had ample time to implement the corrective practices, the state's production had quadrupled to nearly 20 million pounds.

In the absence of pecan rosette, pecan production in the Southeast was, and to some extent continues to be, limited by sunlight, humidity, and in some years, excessive rainfall. A great majority of plant diseases thrive in warm, humid conditions. In the southeastern United States, where heat and humidity saturate the air from June through September, diseases can be a major issue. In rainy years little black specks begin to show up on the leaves and/or nuts of pecan trees. Left unchecked, these little black specks can develop into spots, which can coalesce, turning the entire fruit black, minimizing its size and quality and causing it to fall from the tree. These black velvety-looking spots are known as the most serious disease of pecan, pecan scab.

Otherwise known as *Fusicladium effusum*, pecan scab is a fungal disease that became a major threat to the survival of the pecan industry in the southeastern United States from the Atlantic Coast to East Texas before the development of chemical fungicides. Some of the more popular pecan varieties of the time, such as Delmas, Pabst, and Van Deman, were especially susceptible to the disease, prompting many growers to remove most of these varieties from their orchards. Even Schley, considered the industry standard in quality for much of the twentieth century, was removed from many orchards because of the production limitations brought about by pecan scab. Fortunately, not all pecan varieties have the same degree of susceptibility to pecan scab. Some, such as Elliot, have had a high degree of resistance to the disease for many years.

Pecan scab is a wily fungus capable of adapting in order to overcome a tree's innate defenses against disease. Such adaptability on the part of the pecan scab fungus provides a particular challenge to pecan breeders such as the University of Georgia's Dr. Patrick Conner, who is charged with the achingly slow task of developing new pecan varieties. In a process that moves at glacial speed, each spring Dr. Conner painstakingly removes pollen grains from the male flowers of pecan trees and dusts them onto the female flowers of others in controlled crossing experiments. After harvesting the individual nuts resulting from these crosses, he plants them the following spring and must wait as many as 10 to 15 years to determine whether any of these crosses result in a tree that produces a suitable nut for commercial production. One of the most important traits scientists look for in a pecan is scab resistance. Dr. Conner speeds up the process of screening his crossed trees for scab by growing them for the first year in pots beneath a shade cloth. The trees are regularly watered with overhead irrigation to intentionally favor the development of the disease. This allows Dr. Conner to quickly weed out any individual trees that are particularly susceptible to pecan scab. Those with the desirable level of scab resistance move on to the next phase of testing. Unfortunately, there is no way to determine how long the scab resistance of a particular pecan variety will hold up. Pecan breeders can only evaluate their selections for a few years, release the most promising, and hope for the best.

The development of scab susceptibility in a previously "resistant" pecan variety is something of an arms race between the fungus and the tree, in which, in a humid climate, the fungus has the advantage. To a large degree the outcome is determined by how widely planted the variety becomes. Each variety of pecan tree has its own uniquely adapted race of pecan scab. The Desirable pecan, currently the most popular variety grown in Georgia, was believed to be relatively immune to pecan scab upon its initial release. As thousands of acres of Desirable trees were planted throughout the state, the fungus was, in a sense, "overexposed" to this variety. This allowed the fungus to adapt in such a manner that it overcame disease resistance in Desirable, to the point that the variety is now among the most scab-susceptible grown in the southeastern United States.

Still, the disease is driven almost exclusively by the frequency and duration of wetness on the leaves and nuts.[23] Without wet conditions, the disease will not develop, even on the most susceptible varieties. As early as 1918, C. S. Spooner and C. G. Crittenden, scientists with the Georgia State Board of Entomology, noted, "There is a great seasonal difference in the severity of scab. Very wet seasons promote the growth of the fungus to a great degree. One season has been observed in which practically every Georgia nut in one grove was destroyed by scab before the season was half over. On the other hand, in very dry seasons the loss is hardly noticeable. . . . Between these two extremes the amount of damage will vary according to the season."[24]

Early attempts at controlling pecan scab involved the use of chemicals such as atomic sulfur, lime sulfur, and a popular combination of materials known as Bordeaux mixture.[25] Made of various mixtures of copper sulfate, hydrated lime, and water, Bordeaux mixture has positive effects on plant disease that were discovered by chance in the late 1800s when a French botany professor, Pierre Marie Alexis Millardet of the University of Bordeaux, noticed

(*Left*) **FIGURE 5** Spraying
pecan trees for pecan scab with
a mule-drawn barrel pump sprayer
in the 1930s.

(*Above*) **FIGURE 6** Dusting pecan
trees in the 1930s.

that grapevines growing near the roads around the Bordeaux region were free of a disease called downy mildew, while those farther into the vineyards were affected. Upon inquiring about this phenomenon, he was told that the grapes near the road were sprayed with a mixture of copper sulfate and hydrated lime to deter passersby from grazing on the fruit of these vines.[26] Following Millardet's discovery, Bordeaux mixture was commonly used on grape, fruit, and nut trees to ward off disease. It became the first fungicide to be used worldwide and, in fact, is still used today on certain crops.

Bordeaux mixture and the other materials used were somewhat useful in lowering the incidence of pecan scab, although by today's standards they would be considered relatively ineffective. Now armed with materials that could reduce crop losses to diseases like pecan scab, pecan growers were faced with the challenge of getting these materials up into the canopy of a 50-foot tree. The first applications of Bordeaux mixture to grapevines were made with a broom. In order to get the material into the canopy of orchard crops, orchardists developed a variety of different sprayers, including the barrel pump, which consisted of a 60- to 80-gallon wooden barrel mounted on a mule-drawn wagon. A tower rested above the barrel, on which perched one or two operators bearing hoses to reach the tops of the trees. The operators as well as the mule driver and the mules were freely exposed and were often doused with the materials they were applying. Motor-driven pumps powered by steam, gasoline, or compressed air were used to pressurize the system. Medium-sized trees on smaller acreages could get by with 4- to 6-horsepower pumps, but for larger trees and orchards, 8- to 10-horsepower pumps were used.

As the trees increased to over 20 feet tall, pecan growers observed that the standard spray rods on the ends of the hoses produced a poor stream, too weak to reach the tops of the trees. These were replaced by the spray gun, which could reach 50 to 60 feet above the ground at a pressure of 250 to 300 pounds. Workers were instructed to spray methodically, beginning on the lower limbs and moving right to left and back in short horizontal motions as they worked up to the top of the tree.

One of the earliest sprayers designed for orchard crops was the Bean Orchard Sprayer. John Bean, inventor of the first double-acting force well pump, moved from New England to California's Santa Clara Valley in the early 1880s, seeking recuperation from a health problem. There, he began growing almonds, where he recognized a need for an apparatus that would effectively apply insecticides to his orchards to combat San Jose scale. From this need, Bean developed the first high-pressure, continuous-action spray pump with an air chamber. After Bean exhibited his pump at the 1884 California State Fair, demand for the machine rose to the point that he opened a factory for production in San Jose. In 1901, at 82 years of age, John Bean created a vertical sprayer pump, yielding a higher pressure than any other pump on the market. By 1910, Bean's operation had expanded to a branch factory in Berea, Ohio, and soon, the John Bean Spray Pump Company became the largest manufacturer of orchard spraying equipment in the world. As the company grew, the name was changed to the Food Machinery Corporation, or FMC.[27]

In addition to sprayers, dusters were used to apply pesticides to orchard crops. These machines consisted of a small pump and holding tank mounted to an iron-frame wagon with a large spout at its rear from which the fog of pesticide-laden dust billowed out across the orchard. In 1904, Ernest B. Freeman of Middleport, New York, invented a unique fruit-spraying machine that used carbon dioxide as a carrier gas. The machine was built mainly for the application of Bordeaux mixture, which proved unfortunate for Mr. Freeman when Bordeaux mixture was replaced with calcium polysulfide as a treatment for fruit trees. Calcium polysulfide was not compatible with carbon dioxide, and as a result, Freeman's small company was liquidated. Out of these circumstances, Niagara Sprayer Company was formed, specializing in the production of pesticide dusts and the manufacture of machines to apply them. In 1942, Niagara Sprayer Company was sold to FMC.[28]

These early sprayers and dusters made their way into the South's pecan orchards in the mid-1920s. Bordeaux mixture sprayed onto trees reduced losses to scab, as did copper lime applied by dusting. Early experiments showed that spraying pecan trees could be done for only $1.50 per tree and would result in 95% control, saving an estimated $25.00 per tree. Dusting pecan trees, however, turned out to be a little trickier than spraying them. In order to get the dust to adhere to pecan leaves and nuts, it had to be applied when the leaves were wet, normally after a heavy dew or rain.[29] The problem with this require-ment was that most fungicides are preventive rather than curative, and the wet conditions required to adhere the dust to the plant tissue were the same conditions preferred by pecan scab. This meant that pecan growers had to be ready to dust their trees at a moment's notice when conditions arose in order to prevent scab from getting a good foothold. In most cases, two to three spray applications of Bordeaux mixture gave the same measure of scab control as five dust applications. Because of the strict regimes under which dusters could be effective, these machines did not last long in the pecan industry and were abandoned in favor of sprayers.

The greatest leap forward in pecan pest management came in the years following World War II, with the development of more effective pesticides and air-blast sprayers. The first "fans" used to propel chemicals up into the canopy of orchard crops were airplane propellers used to apply dusts to citrus in Florida in the late 1930s. In 1949, G. W. Daugherty patented a new sprayer using an axial-flow fan rather than an aircraft propeller to jettison the mate-rial into the trees. The axial-flow fan moves the air parallel to the shaft about which the blades rotate, similar to the system in your home air conditioner or car. The air pulled in by the fan is then forced up through a volute where it hits the pesticide coming out of the spray nozzles, which is then forced up into the canopy of the tree.[30] Air-blast sprayers first made their way into pecan orchards in the early 1960s, and by the mid-1970s to early 1980s, all serious pecan producers in the Southeast were using the machines to control insects and pecan scab. Air-blast sprayers continue to be a necessity for commercial production of most popular pecan varieties, particularly in humid areas.

In the early twentieth century, with money hard to come by, the 8 to 10 years required before pecan trees began bearing was a particularly difficult ob-stacle for prospective pecan farmers to overcome. Because pecan trees require such a long time to begin bearing on a profitable scale, many early pecan farm-ers planted a variety of crops between their tree rows in order to make the most efficient use of the land. Many interplanted cotton, corn, Irish potatoes, sweet potatoes, velvet beans, soybeans, peanuts, or vegetables for income off the land. A seminal text on pecan production called *Pecan Growing*, written by H. P. Stuckey, director of the Georgia Agricultural Experiment Station, and E. J. Kyle, dean of the School of Agriculture at Texas A&M, gave this advice on making best use of the land: "Two crops can be grown in the orchard in one season by first planting an early maturing sort, such as Irish potatoes, or any of the early vegetables that can be harvested by May 20 to the 10th of June, and then followed with some late maturing kind, such as sweet potatoes, peanuts, tomatoes, eggplant, peppers, or in some cases cotton."[31] This way of thinking regarding the use of the land to its fullest potential was not only an ideal thought at the time, but a necessity for many farmers.

During a discussion at the Fifth Annual Convention of the National Nut Growers Association in 1906, Professor H. E. Van Deman stated that "cotton is one of the best crops grown in the pecan orchard," along with corn and cowpeas. "Never, never, never put any small grain in an orchard," Van Deman continued. "The next thing to a fire [in an orchard] is an oat crop."[32]

During the same conference, J. B. Wight suggested, "Legumes are the best of all [crops to be grown in an orchard] in that they gather their supply of nitrogen. . . . Peas are the most convenient legume to apply. Velvet beans are also excellent, but are objectionable among bearing trees in that they are very much in the way in gathering the nuts."[33]

Even kudzu, "the weed that ate the South," as it has been called, was once promoted as a cover crop in pecan orchards. In the 1920s, J. Slater Wight grew the aggressive vine between his tree rows and cut them once a year for hay, suggesting, "Kudzu is the best summer legume to grow in the pecan orchard for adding organic matter to the soil."[34] Kudzu made its debut at the same Philadelphia Centennial Exhibition where Hubert Bonzano introduced the world to Centennial, the first vegetatively propagated pecan. The kudzu vine with its fragrant, purple flowers grew over and partially covered the Japanese Pavilion at the exhibition. Because it lacked cold tolerance, kudzu never became a problem above the Mason-Dixon Line. But in the South, it was planted as an ornamental for screening and for its pleasantly scented blooms. By the 1930s, the US Soil Conservation Service was advocating kudzu as a means of alleviating the effects of soil erosion across the South. During the Depression, the Civilian Conservation Corps planted 73 million kudzu seedlings along roadsides and eroded farms. By the 1940s Kudzu Clubs were being formed to honor and promote the vine's virtues.[35] Finally branded a weed in 1967, kudzu would eventually cover over seven million acres of the southern landscape with its iconic dense tangles of vine that overtook anything in its path.

In some areas, particularly in central Georgia near Fort Valley, peaches were interplanted with pecans as a viable crop until the pecan trees could begin to bear commercial crops. Peaches begin bearing approximately the third year after planting and were well suited to this arrangement. In a 1922 report to the National Nut Growers Association, W. H. Harris of Fort Valley, Georgia, suggested that "from a cultural standpoint, there is no difficulty in intercropping pecans with peaches. They ripen their fruits several months apart, usually with heavy rains intervening so that they conflict very little. They require the same cultivation, and my experience has been that spraying the peaches is most beneficial to the pecans."[36]

Harris described two methods of interplanting. One involved simply planting a pecan tree instead of a peach tree at every third tree space in every third row. By placing his peach trees 20 feet apart, he would be left with a 60 foot × 60 foot arrangement of pecan trees. Another plan, less preferred by Harris, was to set out a row of pecan trees followed by two rows of peaches. According to Harris, this made it difficult to harrow and drained the soil of water and nutrients to the point that both the peach and pecan trees were dwarfed. Others got around this problem by planting only a single row of peach trees between the pecan rows. A few remnants of the peach/pecan interplanting arrangement persist in the Fort Valley area today.

GEORGIA

Pecan producers have always led the growth of the pecan industry by pioneering methods and practices for bending Mother Nature to their will (within certain limits), allowing them to do things that the "experts" said couldn't be done. At the age of 13, Fred Voigt moved with his family from Atlanta to the small rural community called Forks of the River in Pierce County, near Waycross, Georgia. When the Voigts first arrived on their new farm in 1914, it had a few seedling pecans standing here and there that bore small but delicious nuts. At that time, the forests of the area were full of pine trees 24 to 30 inches in diameter. The economy of southeast Georgia depended on these trees, which were tapped for turpentine and cut for timber. When the woods were flooded from heavy rains that overflowed the Satilla River, locals went out into the swamps and cut timber with a crosscut saw while standing in waist-deep water. Families worked together, forming rafts from the large logs held together with long iron pins connected to rings that were threaded with heavy rope. The logs were then rafted downstream to a sawmill in Brantley County.

The Voigts made syrup from sugarcane and grew vegetables, hay, pork, and beef, all of which were loaded onto wagons and transported to town for sale each Thursday. But, of all the bounty of nature surrounding Fred Voigt on this south Georgia farm, it was the pecan tree that would draw his interest and become a lifelong love. It would take the unlikely influence of two men, one from Pennsylvania and the other from Illinois, to spark the young Georgia farm boy's interest in pecans.

A retired professor from Pennsylvania, Albert Clark Snedeker, was a dynamic promoter of pecans in the early twentieth century. In his retirement, Snedeker became a nurseryman, and it was from this man that Voigt learned to bud, graft, dig, and pack nursery stock while working in Snedeker's nursery. Snedeker, who served for a time as president of the Georgia-Florida Pecan Growers Association, championed the planting of pecans throughout southeast Georgia and into the Carolinas. R. Lloyd Scott also operated a commercial nursery in Blackshear. A retired horticulturist from the University of Illinois, Scott taught Voigt the art of bark grafting, which the latter used to establish a successful business grafting the many seedling trees that were so abundant on the area's small farms over to Schley, Stuart, Pabst, and Delmas.

Once on his own, Voigt acquired a 215-acre farm in Ware County in 1933. Only 35 acres of the farm were in cultivation and the rest of the low land was covered by pine timber and cypress ponds. Most of the land was wet with a high water table, land not generally conducive to pecan production. Yet in 1938, Voigt planted his first orchard of five acres using nursery stock he had grown himself. A few years later he added 30 more acres. In 1959, Voigt decided to clear the remaining land on the farm and began planting it to pecan trees, eventually planting a total of 200 acres.

Voigt solved the problem presented by the low, wet land with the creation of a drainage system that shaped the land into gentle, sloping ridges and valleys. Trees were planted on top of the ridges, and the valleys were sloped into a drainage canal that moved the water away from the field. This method prevented waterlogging and ensured adequate moisture from the high water table, creating an environment that closely mimicked that of the pecan tree's native habitat. The trees responded accordingly. Soon Voigt's orchards were generating excellent yields while others in the area struggled. This success was perhaps best demonstrated during the drought of 1968, in which pecan production was drastically reduced in much of Georgia, while Voigt produced large, well-filled pecans. By creating a growing system to which the tree was adapted rather than trying to force the tree to produce in undesirable conditions, Fred Voigt became one of the first, if not the first, to grow pecans on a commercial scale on the low, wet, sandy soils on the edge of the vast Okefenokee Swamp.

Through his innovative and curious mind, Fred Voigt became a leader in the pecan industry. Ray Worley, longtime pecan researcher at the University of Georgia, once wrote that "several PhD's received their practical pecan training" under Voigt's able supervision. He generously offered up his own orchards to researchers for study in order to find answers to problems that plagued the pecan industry, particularly regarding fertilization.[37] The answers generated on his farm turned Fred Voigt into something of a local legend in his own time, and he was often the first person people in the area turned to when they had questions about pecans.

Although Georgia came to lead the nation in pecan production by 1950, it was 1963 before the state would break the 100-million-pound mark in production. Following that year, Georgia production bounced up and down with the peaks and valleys of the alternate bearing cycle as its once-vigorous young trees began to mature and their management became more difficult with increasing size and age, requiring greater care. In the 1980s, Georgia's pecan production would soar, falling below 100 million pounds only three times from 1980 to 1990. Much of the credit for this dramatic increase in pecan production lies with University of Georgia extension specialists Tom Crocker, H. C. Ellis, and Paul Bertrand, who championed a "package" approach to pecan production, including fertilization, insect and disease management, irrigation, and sunlight management. Many of the original pecan orchards in Georgia had become very crowded by this time as the trees grew and competed with each other for sunlight, water, and nutrients. The resulting production was limited, as trees were unable to perform the physiological processes necessary to produce a crop in the limited sunlight they received.

The trio of Crocker, Ellis, and Bertrand promoted removal of trees to open up orchards and replacement of old, nonproductive varieties with those that had better characteristics. This strategy, they taught, would not only improve the trees' production but would make the management of diseases and insects much easier since the trees would be under less stress. The practices they promoted helped keep the trees as healthy and stress-free as possible so that they could exert their energy toward nut production. While many of these practices were already being put into use by more-progressive orchard managers in the Albany area, it was Crocker, Bertrand, and Ellis who took this message to the masses and improved pecan production throughout the state, giving the Georgia pecan industry new life to carry forward into the next century.

TEXAS

While the volume of pecans produced for commercial sale in Georgia exceeds that of any other state in the United States, Texas has more pecan trees than virtually any other place on earth. The trees grow naturally along most of the 8,000 miles of rivers coursing throughout the state, particularly along the Red, Sabine, Trinity, Neches, Nueces, Guadalupe, Colorado, and Brazos Rivers and their tributaries. It has been estimated that there are from 600,000 to one million acres of native pecan trees in Texas because the tree grows wild in 150 of the state's 254 counties.[38] Only a fraction of the native acreage is managed consistently for nut production. In addition, there are about 67,500 acres of improved pecan varieties growing in orchards across the state. Nearly 80 percent of the counties in Texas have commercial pecan orchards.

The center of Texas pecan production in the early years was San Saba, in central Texas. Today San Saba is known for two things: it's the home of actor Tommy Lee Jones and, more significantly, pecans. Located in the heart of the Texas Hill Country, San Saba is known as "The Pecan Capital of the World." It arrives at this distinction primarily through its abundance of native pecan trees. Its proximity to the Colorado and San Saba Rivers, whose banks are lined with wild pecan trees, makes San Saba a perfect site for the tree. Pecans have been a cash crop in the San Saba area since as early as 1857. From its ancient native trees, the county produced 3.5 million pounds of pecans in 1919, enough for shipping out 60 carloads by rail. No other state in the union produced half as many pecans in 1919 as San Saba County. Today, pecan trees fill the orchards, yards, and landscapes of the region.[39] Normal production for the county is two to five million pounds. As much as 75 percent of these are native pecans.[40]

The state's first native-born governor, James Hogg, is said to have famously remarked on his deathbed: "I want no monument of stone or marble, but plant at my head a pecan tree and at my feet an old fashioned walnut tree. And when these trees shall bear, let the pecans and walnuts be given out among the people of Texas, so that they may plant them and make Texas a land of trees." In fact, the former governor made this remark not on his deathbed, but while he and his daughter Ima were visiting the home of Hogg's law partner, Frank Jones, on the night before his death. Late the next morning, March 2, 1906—Texas Independence Day—Hogg died in his sleep. Shortly after his burial in Austin's Oakwood Cemetery, the fledgling Texas Nut Growers Association planted Russell and Stuart pecan trees at the governor's head and a native black walnut at his feet. As the trees began to bear fruit, the Department of Horticulture at Texas A&M University gathered the nuts and distributed them to individuals, schools, and organizations around the state per Governor Hogg's wishes. A resolution by the Texas legislature in 1919 declared the pecan the State Tree of Texas. In time, both the Russell pecan and the black walnut at Hogg's grave site died. In 1969, in observance of Arbor Day, the Texas Pecan Growers Association planted a Choctaw pecan tree to replace the lost Russell tree.[41]

Needless to say, pecans are big in Texas, which, of course, has always had the reputation of doing things big. One of the earliest pecan plantings of significant size was the 1885 planting of 400 acres at Brownwood, Texas, by F. A. Swinden.[42] Up to that time in Texas, most trees were maintained exclusively along streams, often in single file, unless widespread lowlands allowed for more-dispersed groves. Swinden chose large, soft-shelled nuts as the seeds from which to grow 11,000 trees. Swinden's orchard encountered numerous problems. Not long after the trees began growing, they were attacked by wood lice, which destroyed nearly half of the planting. Swinden then interplanted the orchard with cotton, corn, and alfalfa, the latter of which sapped so much water from the soil that more than 300 trees were killed.[43]

Pecans are so intimately associated with Texas that there is probably no discernible starting point to the pecan industry in the state. The most apparent origin of visible effort in the commercial cultivation of pecans in Texas would have been the propagation efforts of E. E. Risien described previously. Risien's work allowed the pecan producer to grow the pecans of his own choosing rather than simply take what nature supplied. Still, there would be much work ahead to develop the pecan into a respectable crop.

The greatest obstacle facing the development of a commercial pecan industry in Texas early on was the same one that plagued most of the South—cotton. Many of the early pioneers of this region failed to recognize the value of pecans beyond the readily available timber. Massive acreages of pecan-filled river-bottom land were cleared to make way for cotton. The introduction to a 1908 publication by the Texas Department of Agriculture, entitled *Pecans and Other Nuts in Texas*, states, "The Texas pecan has been treated like the other natural resources of the State—exploited, wasted, destroyed. . . . Our fertile lands have been robbed of their productivity by the one crop system; our forests have been cut down for the best timber, and no provisions made for future supply; the pecan has fallen before the axe with its fellows—the elm, the ash, the oak—in the effort to make way for more cotton."[44]

These sentiments were echoed by E. W. Kirkpatrick of McKinney, Texas: "We have hurt ourselves and our soil raising cotton . . . and dragged our women and children down . . . and reduced them to almost slaves to pick cotton and sell it for 5 cents a pound. . . . We have pursued the wrong course in neglecting the pecan . . . it excels everything." The difficulty of establishing the pecan as a commodity in Texas was, in part, a product of the abundance of pecan trees in the state. Most people saw no need to plant trees and tie up land with a tree that already grew wild all over the state. It is truly remarkable that the pioneers of the US pecan industry, not only in Texas but throughout the South, were able to gain any foothold whatsoever for the industry in a climate so intolerant of change. Considering the incredible genetic diversity in the native pecan population today, it is startling to consider what potential varieties may have been lost through the early clearing of pecan bottomlands.

In an effort to raise the pecan to a more respected plane in the state, a group of individuals, with Kirkpatrick serving as president, established the Texas Nut Growers Association in 1906 to "provide for meetings of Texas Nut Growers, and others who may be interested, for the discussion of the various problems that affect the welfare of practical nut growers, and to encourage the growing of better varieties of the Texas pecans and other important nut trees. To assist in carrying out the last request of our beloved ex-governor James Stephen Hogg."

Indeed, the planting of the trees at Governor Hogg's grave site did much to generate interest for and popularize pecans in the state. Because of the large number of native pecan trees, the process of top-working native trees to improved varieties became a popular practice in Texas and Oklahoma throughout the early twentieth century. Top-working involves grafting or budding an existing tree to another variety after severely pruning the trees and allowing them to resprout. These efforts were led largely by Halkert A. Halbert, a pecan producer from Coleman, Texas. Born in Mississippi, Halbert moved with his parents to Texas in 1849. Well educated, Halbert taught school and later moved to Corsicana, Texas, where he practiced law and published a newspaper, the *Corsicana Light*.

Like many pecan men, H. A. Halbert was something of a character. He was the author of an almanac that provided weather forecasts inspired by the heart of an oak tree. Despite the source of Halbert's prognostications, many people depended on his almanac for guidance in the planting and growing of crops. When his health began to fail, Halbert moved his family to the West Texas town of Coleman in 1886. He eventually purchased a 350-acre farm near Coleman on which grew many pecan trees. One of these, bearing a thin shell and of good quality, Halbert christened the Halbert pecan, which he set about to promote and propagate with his top-working technique.[45]

Halbert traveled extensively throughout Texas from 1886 until his death in 1926, continually top-working pecan trees, mostly to Halbert. He often went to the extreme in preparing trees for top-working: "The entire top from six to twelve feet from the ground must be taken off. . . . Do not cut below six feet for fear the cattle may rub off the buds. . . . The pecan is very tenacious to life, and there is no danger of killing it unless too old to be budded. If it is too old for that purpose, you had better apply the axe to the root of the tree and let some sapling it overshadows have the sunshine."[46]

The *Seguin Enterprise* wrote of Halbert's top-working activities in the Seguin area in 1907, "H.A. Halbert of Coleman has been here in Seguin several days, budding pecan trees. . . . Mr. Halbert is an experienced pecan man and says the buds will bear the third year and give paying crop the fourth year. He uses the paper shell Halbert pecan and guarantees them to produce every year. Mr. Halbert insures his work and continues to make sure you have a living tree. He puts buds on the old trees. . . . Mr. Halbert has a tree that gave him a revenue of $253.75 for one year."

Halbert had heavy, dark eyebrows and in his older years, a graying goatee. Aside from his horticultural skills, Halbert had a way with words, which led him to promote the virtues of the Halbert pecan with the following rhyme:

Halbert pecans will never fail

Unless destroyed by pelting hail

No drouth [*sic*] nor late nor early frost

Will cause a crop to be totally lost

So great the yield the trees can spare

Much fruit to worms and leave a share

The nuts, not largest Jumbo size

But sixty best with pound comprise

So thin the shell, you readily tell

They are the truest paper shell

The nuts so sweet and full of meat

That most complete, are others beat

Now don't be small and eat them all

Let future folks "rise up and call

You blessed," and thank with a grateful Heart

For planting them, the smallest part

Not only will these nuts come true

And bear the best of food for you

If care is taken, in years a few

But generations continue to do[47]

Although age crept up on Halbert, it never slowed him down. He died at the age of 78 doing what he loved best—top-working a pecan tree.

Texas made many of the first early strides in the application of science to the art of pecan growing. Dr. E. J. Kyle served as dean of the Texas A&M School of Agriculture from 1900 to 1948 and developed the first academic course in pecan culture in 1911. Kyle continued to teach the course until his retirement in 1948. Sustained through the years by Benton Storey, George Ray McEachern, Larry Stein, and Monte Nesbitt, the course continues to this day as an important part of the curriculum in the Texas A&M College of Agriculture and as a valuable source of information for the future leaders of the pecan industry in Texas.[48] Kyle's other great contribution to the pecan industry was his coauthorship, along with University of Georgia Agricultural Experiment Station director H. P. Stuckey, of the first true text on pecan culture, *Pecan Growing*, in 1925.

The first formal "student" of pecans was G. H. Blackmon, who graduated from Texas A&M in 1910 with a bachelor's degree in horticulture. During his junior and senior years, Blackmon focused mostly on the propagation and variety evaluation of pecans. He worked for a few years as a horticulture instructor at Texas A&M and then as a nurseryman with Waxahachie Nursery Company in Waxahachie, Texas, before being hired as a horticulturist with the University of Florida, where he would go on to make many important contributions to the pecan industry of the Southeast.[49]

Another of Texas's early pecan pioneers was J. A. Evans, a nurseryman and one of the first extension specialists for pecans with the University of Texas at Arlington. Evans was hired in 1917 and, like Halbert, promoted the top-working process. Through his efforts, Evans established islands of interest in pecan production in nearly every section of the state. Some have even suggested that Evans's work marked the beginning of the commercial pecan industry in Texas.[50]

James Henry Burkett was hired as chief of the Division of Edible Nuts of the Texas State Department of Agriculture in 1916. Originally from Cannon County, Tennessee, Burkett moved with his parents from their farm on the site of the Battle of Murfreesboro to Mills County, Texas, in 1865. They eventually settled in Burnet County, where Burkett's father died. At the time, although J. H. Burkett was only 17 years old, he quit school with what he regarded as only a third-grade education. Burkett eventually bought a farm in Callahan County in December 1899 near Battle Fish Creek, just east of Putnam, and began working the land with his young bride and new family.[51]

In yet another remarkable turn of fate, Burkett's two young sons, Omar and Joe, went squirrel hunting in the fall of 1900 and discovered a few large, round nuts of high quality in a squirrel's nest along the south bluff bank of Battle Fish Creek. Burkett urged the boys to show him where they found the nuts, and after searching the forested bottom, the boys were able to locate the tree

FIGURE 7 The J. H. Burkett family. Mr. Burkett (center) was one of the first government pecan specialists, and the namesake and developer of the Burkett pecan, one of the most popular early pecan varieties in Texas. Two of Burkett's sons discovered the nuts of the original tree in a squirrel's nest along the banks of Battlefish Creek in Callahan County, Texas.

that had produced the nuts. It grew on land owned by W. A. Orr and was protected on three sides by live oak and mesquite trees. Between the tree and the river was a tall elm, which protected it from the nearby river's eroding waters.

Burkett immediately began attempts at budding the tree the following spring but had no success until finally, in 1903, he successfully transferred two buds of the original tree to a seedling.[52] The budded tree produced nuts in 1905, and from here Burkett went on a quest to acquire as much knowledge about the pecan as he could. This led to recognition of his expertise and his appointment as nut specialist within the Texas Department of Agriculture. Burkett began selling graftwood of the Burkett pecan in 1911 and by the 1930s, the Burkett pecan was the most popular variety in Texas because of its size and quality. Sadly, the original parent Burkett tree growing along the banks of Battle Fish Creek was destroyed in 1910 as a result of a dispute over the propagation rights of the tree. The original tree budded by Burkett persisted under protection by the state of Texas in Callahan County on the north edge of I-20, about one-half mile east of FM 880 near the Eastland County line, until 1993, when it was killed by lightning.

Certainly one of the first government pecan specialists, if not the first, Burkett made many contributions to the development of the pecan as a viable crop in Texas, but aside from the Burkett pecan, perhaps his greatest contribution was the expansion of Texas pecan production into West Texas. The overriding factor that made this expansion possible in such a desert region was the use of irrigation.

Burkett's work to expand pecan production into the dry desert west was made possible by the observations of Gilbert Onderdonk, a pioneer horticulturist known as the "Nurseryman of Mission Valley." Onderdonk was another fascinating character in the story of pecans and Texas's early history. Onderdonk came to Texas from New York in 1851 seeking an environment more conducive to his health. He loved everything about Texas. His health quickly recovered and he became a self-taught horticulturist of extraordinary skill and knowledge. He served as a Confederate soldier during the Civil War and is credited with almost single-handedly developing the state's fruit production. He gained international recognition throughout Europe for his horticultural expertise.

Onderdonk was asked by David Fairchild of the US Department of Agriculture's Office of Plant Introduction to serve as a special agent for fruit crops.[53] He would explore the plant life of Mexico and report his horticultural findings. In 1911, Onderdonk published "Pomological Possibilities of Texas," in which he reported on the thousands of 200-year-old pecan trees growing in Bustamante, Mexico, with irrigation. He also reported on the minimizing of alternate bearing with irrigation at Saltillo and Atoyac. At that time, pecans growing on the banks of Texas rivers were said to produce a crop only once every three years, and on the dry uplands of South Texas pecans were reported to rarely bear a nut.[54]

Irrigation would later prove to be one of the most valuable innovations in producing pecans throughout the nation, allowing production to make yet another leap forward. In an address to the National Nut Growers Association in 1921 entitled "The Future of Pecans," R. J. Bacon speculated on the potential yields of the pecan tree. "In the very near future of pecans, I am convinced we will have very heavy yields per tree in comparison to any averages that we have been able to maintain so far." The key, Bacon realized, in addition to proper fertilization, was irrigation. "I am familiar," Bacon said, "with a swimming pool built between two pecan trees. It is a crude construction with ingot iron sides and sand and clay bottom. These trees have begun some very wonderful bearing with no apparent cause other than having about 1/3 of their root system constantly underwater. I am familiar with trees next to a horse lot, that are green in dry weather and year in and year out bear tremendous crops. And if the conditions could not be duplicated any other way, these increased yields . . . would authorize the cost of horse lots and swimming pools all over the grove."[55] Fortunately, more efficient means of irrigating pecan trees were developed. Through the use of irrigation, pecan production in the Desert Southwest went from an impossibility to a profitable endeavor, and in the southeastern United States, production consistency was enhanced and yields were increased by 150 percent.

The oldest method of irrigation is flood irrigation, whereby water is directed down a series of canals or aqueducts from the water source to the fields in which it is needed. This method is nearly as old as agriculture itself and was practiced in Egypt as early as 5000 BC, when King Menes diverted water from the Nile River to adjoining arable lands. Subsequently, this technology was independently developed throughout the civilizations of the world before recorded history. In fact, some believe that irrigation made civilization itself possible.[56]

In the southwestern United States and Mexico, irrigation canals have been used since the time of Columbus. During the Spanish conquest over

300 years ago, the orchards of Aguascalientes in central Mexico were famous for their peaches, grapes, figs, and pecans. As Europeans moved into the Rio Grande Valley, similar irrigation systems were developed. Particularly in the arid regions of Texas, New Mexico, Arizona, and California, irrigation has made agriculture possible. This miracle includes the production of pecans in those regions, including the trees Gilbert Onderdonk observed in Bustamente.

Flood irrigation of pecans is still used today in the southwestern United States and is best suited to relatively heavy soils with little sloping. Border dikes are usually built to hold the water in a basin, ponding the water across the surface of the orchard. Because of the large amount of water covering the surface of the orchard when flooded, flood irrigation may appear to be a very wasteful or inefficient method of providing water to the trees. Surprisingly, with the right soil type, slope, and flow rate, it can be very efficient.

Flood irrigation provides a very uniform wetted area, which is ideal for mature trees with large root zones in relatively heavy soil. Soil is indeed the most valuable link in the efficient use of flood irrigation, for which soil texture, structure, profile, and hydraulics are key. Most of the disadvantages of flood irrigation that may occur through soil type, slope, and so forth can be managed by controlling the grade with land leveling, and by managing the width of border dikes, the length of the water run, or the amount of water available at any given time. One advantage of flood irrigation is the low initial cost, which is limited to water supply development and land preparation. Flood irrigation systems do, however, require considerable skill and labor.

In addition to flood irrigation with border dikes, furrow irrigation systems are used in which furrows are constructed parallel to the tree rows. Because these are normally placed near the trees, tractors in the orchard can damage tree roots when operating too close to the furrows. At harvest, these furrows are smoothed out to prevent interference with harvest operations.

Until the 1980s, many people scoffed at the thought of irrigated pecan orchards in the southeastern United States. Pecans were considered to be tolerant of drought because of their deep root system. In addition, the humid Southeast was thought to receive plenty of rainfall to accommodate the water requirements of pecan trees in most years. However, pecan producers and scientists began to realize that rainfall was not always dependable and was usually in short supply at the time of year it was most needed. August and September are especially critical times for adequate soil moisture as the pecan kernels fill the nuts. Yet these are also historically the driest months of the year in the region. Although the pecan tree is deep rooted and is capable of reaching water at considerable soil depth in order to keep the tree alive, the shallow feeder roots work to satisfy much of the moisture needs of the developing crop, and in dry periods these roots become nonfunctional. As a result, supplemental irrigation has proven to be highly beneficial to southeastern pecan production, accounting for a little over half the water used to produce the crop.

Serious speculation about the benefits of irrigating commercial pecan orchards in the Southeast began in the late 1940s. The rolling land of the region was not conducive to flood irrigation, so other methods had to be developed. The first mention of irrigation in the pecan literature comes from a 1947 article in the 40th *Proceedings of the Southeastern Pecan Growers Association* in which Everett Davis, an irrigation engineer with the University of Georgia, describes a portable revolving sprinkler system that he calls "the first irrigation system of any type and of any consequence" on the L. R. Barber farm in Colquitt County, Georgia.[57] The article described Barber's conviction that "irrigation of his pecans . . . would make a difference of $10,000 in his income from the crop." This conviction was based on the potential of late-season irrigation to solve the problem of poor quality and quantity resulting from dry periods occurring late in the season. Although we don't know whether or not the system in question truly increased Barber's pecan income by $10,000, we certainly know that irrigating pecans during this time improves kernel quality and production.

Scientific experiments on pecan irrigation began in the mid-1950s. One of the earliest studies occurred in Texas, where A. O. Alben found yields to be highest from well-irrigated trees.[58] Since then, numerous studies from all over the pecan belt have consistently reported increased growth, production, and yield of pecan from irrigation.

Portable sprinkler irrigation systems soon gave way to more permanent "solid-set" irrigation systems, the first of which began to show up across the southeastern pecan belt from Georgia to Texas beginning in the 1960s. Solid-set systems consist of permanent aboveground sprinklers set on risers about two feet off the ground. These sprinklers are connected by a series of buried PVC lines running throughout the orchard. Solid-set sprinklers provide coverage of a large surface area but often at a great financial cost because of the energy required to pump such a high volume of water. In addition, solid-set systems are probably one of the most inefficient means of applying water to pecan trees since much of the water is lost to evaporation or is used by grass and other vegetation growing between the tree rows. As a result, very few solid-set sprinkler irrigation systems are being installed in pecan orchards today.

In Georgia, pecan orchards were seldom irrigated prior to 1970. By 1987 about 55 percent of commercial pecan orchards were irrigated. By 1997, nearly 68 percent of commercial pecans in Georgia were irrigated. About half of these orchards were irrigated using drip irrigation.[59] It is difficult to get a handle on the exact percentage of irrigated pecan acreage in Georgia today, but estimates place it at over 80 percent and growing.

By the early 1970s the commercial planting of pecans had accelerated in arid central and West Texas, particularly near El Paso. At this time, drip irrigation came into use and many growers found it a better option for pecan irrigation.[60] Drip irrigation has its origins in early 1930s Israel, where a water engineer named Simcha Blass was drawn to a large tree growing in a yard with no water. Blass noticed a small wet area on the soil surface near the tree. Digging below the surface, Blass found the source of the wet spot, a leaky coupling. More intriguing to the young engineer was the expanding, onion-shaped area of moist soil underground reaching the tree's root zone. This sparked the concept of drip irrigation in Blass's mind. Understanding the potential for this discovery, Blass set about turning his idea into a product that has revolutionized the irrigation of horticultural crops, including pecans.

After leaving government service in 1956, Blass opened a private engineering office and worked with his son Yeshayahu on the practical development of drip irrigation. To do this, Blass developed a way to release water through larger and longer passageways (rather than tiny holes) by using friction to slow water inside a plastic emitter. The larger passageways prevented the blocking of tiny holes by very small particles. The first system of this type was used experimentally in 1959 and by the early 1960s, Blass had developed and patented the first practical surface drip irrigation emitter.[61] By 1973, there were approximately 25,000 acres of pecans on drip irrigation in the United States.[62]

The water delivery system is made up of a network of plastic tubing with lines terminating at an emitter. Drip systems in pecan orchards usually consist of one larger main line off which usually two smaller lines run parallel to each tree row with 6 to 16 emitters per tree. There are numerous advantages to drip irrigation but the most obvious is the savings in water used. Water is applied slowly through each emitter at low rates of one to two gallons per hour under low pressure so that little is lost by evaporation or wasted in areas where it is not wanted. Water-soluble fertilizers can also be applied through the drip system, allowing for highly efficient and cost-effective methods of fertilization. The system is easily automated and its operation costs are relatively low. Because of its high efficiency and the low cost of materials and operation, drip irrigation caught on quickly throughout the nation as a method of irrigating pecans.

Microsprinkler irrigation came into use by pecan producers during the 1990s. Similar to drip irrigation systems in that they use a low volume of water, microsprinklers deliver small droplets, have low-flow application rates, and distribute the water over a larger area than does drip irrigation, and some feel that this aids in the establishment of tree roots. Microspinklers operate at low pressure and help maintain low water tension in the soil, allowing plants to remove water from the soil more easily, thanks to the low-flow application rate.

Some microsprinklers operate with a spinner, which is typically seated on a nozzle. When water passes through the nozzle it spins the spinner, which throws the water in a full circle to the diameter determined by the flow rate and pressure. Other microsprinklers, often termed microjets, have no moving parts and distribute the water in a manner very similar to that of the microsprayers. The water passes through the nozzle in a straight stream, hitting the spreader, which has from 6 to 16 grooves that deflect the water into narrow, multijet streams.

Many now recognize irrigation as the single most important component in growing pecans. It is now hard to fathom, in an age when it is not recommended to even grow pecans without some form of irrigation, how the pecan industry was able to get off the ground without it. There is no other practice that will do more good for pecan production and pay back on its investment as quickly as irrigating pecan trees.

The problem of soil salinity following irrigation became an issue in arid pecan-producing regions. Many soils in arid regions contain naturally high concentrations of salt. In general, these areas do not receive enough precipitation to weather the soils enough to leach out the salt. Irrigation water and fertilizers may contain salts that contribute further to the problem. Compacted soils, heavy clay soils, or sodium problems may prevent downward movement of water and salts, making control of soil salinity difficult. Adequate soil drainage is needed to allow leaching of salts below the root zone of the trees if soil salinity is to be managed. Sandy soils have larger pores that allow for more rapid drainage. Those with a high clay content may have poor drainage and lead to the accumulation of salt in the root zone.

The most serious problem caused by soil salinity is the reduction in usable water available to the tree. Pecan trees grown in salty soils will experience moisture stress sooner than those grown in better soils. Their growth rate, nut size, and yield will be reduced. Branch dieback can occur, and in severe cases, trees may die. Soils high in salt may also contain toxic levels of sodium, lithium, boron, and/or chloride, which can poison the tree. There are no magic bullets that can directly control soil salinity. The key is adequate soil drainage, which allows salts to leach down below the tree's root zone. Maintaining soil drainage and providing good irrigation management are the major means by which growers manage soil salinity in their orchards.

Before the controversy over migrant farm workers and illegal immigration laws surfaced to disrupt American agriculture in the twenty-first century, similar issues plagued one segment of the pecan industry early in the 1900s. Perhaps more than any other city, San Antonio occupies a historical place in the lore of Texas pecans. During the 1880s, San Antonio was the center of the pecan trade. Half of the commercial pecans in Texas grew within 250 miles of the city at that time. Over 1,250,000 pounds of pecans were marketed in San Antonio in 1880. Central Texas frontiersmen sold bags of pecans in San Antonio for a nickel per pound just as commercial orchards came into being, and by 1882 Gustav Duerler Sr. was hiring Mexican workers to crack the nuts with railroad spikes and shipping the shelled kernels to the East Coast. Here, nearly 48 percent of the national pecan crop was shelled. Workers were paid $2.50 per day or six cents per pound of shelled kernels for their labor.

By the 1920s, improved pecan varieties grown in Georgia and other areas of the Southeast commanded the highest prices, but they were often shipped as far away as Chicago or Saint Louis for shelling, requiring pecan buyers to pay a little extra for the product. Seedling pecans, usually much smaller than their improved brethren, were used in ice cream and confections rather than as stand-alone treats, as were many improved pecan varieties. The capital of seedling pecan production was and remains Texas. Texas shelling operators learned early on that by shelling native or seedling pecans in San Antonio and shipping only the meats or kernels to their customers, they could reduce their expenses by nearly two-thirds.

Like other manufacturers, pecan-shelling operators required mechanization to streamline their business. The first commercial-scale cracking of pecans began with the Barnhart Mercantile Company of Saint Louis in 1884. In this operation, a hammer with a block of lead as a base was used for cracking the nuts. The first lever-operated pecan-cracking machine was developed by Robert E. Woodson in 1889.

By 1914, Woodson had invented the first power-driven cracker. While useful for nuts of uniform size, the rudimentary machines often cracked and damaged the seedling pecans, which came through the shelling plant in a plethora of shapes and sizes. Because consumers paid a premium price for unbroken kernels, pecan-shelling operators in San Antonio used hand shellers through the 1930s in order to process them.

Gustav Anton Duerler was born in Saint Gallen, Switzerland, in 1841 and immigrated with his family to Texas in 1849, settling in San Antonio.

In his youth, Duerler was apprenticed to a printer and worked in that field and in clerical positions before serving in the Confederate Army during the Civil War. He returned to San Antonio after the war and established a confectionery that eventually grew into the Duerler Manufacturing Company. By 1928, Duerler's cracking and grading operation was completely automated and he employed 1,000 Mexican women and girls to shell and package the pecan meats for shipping. Duerler was becoming one of the industry's most powerful men and dominated the pecan-shelling industry by 1930. Duerler and the other shelling factories around San Antonio controlled both the supply and the price of pecans. However, a competitor with a different mode of operation in mind would soon present a formidable challenge to Duerler.

Julius Seligmann had by all accounts done very little to distinguish himself when, in 1926, he inherited his father's land near Seguin and, with $50,000 cash, founded the Southern Pecan Shelling Company. In order to compete with big shellers like Duerler, Seligmann decided to use the readily available cheap labor in San Antonio rather than mechanize to process pecans. He would develop, in a sense, sweatshop pecan-shelling operations.

Under the continued booming economy of the 1920s, Seligmann's throwback model of a shelling operation would most likely have failed. Its brief success was due in large part to the Great Depression. When the economy failed, mechanized shellers found it difficult to keep up with the continuous maintenance and upgrades on their machinery. Seligmann, on the other hand, had few capital expenses, and he could still ensure a profit simply by slashing the wages of his workers. Duerler's shelling operation soon failed amid the growing calamity of the Great Depression.

With Duerler and other major pecan shellers out of the way, Seligmann soon developed a monopoly over the pecan industry that allowed him to set wages for all operators in San Antonio. The city's Mexican population was soon forced into low-paying jobs. During the winter of 1933 to 1934, San Antonio's shelling operators employed 20,000 people in 400 facilities across the city. By 1935, Southern Pecan Shelling Company was shelling approximately one-third of all the pecans in the United States. In 1936, with the rest of the country in the depths of the Depression, Southern Pecan grossed over $3 million profit, while its laborers were among the lowest paid in the country.

Amid great suffering, the Mexican workers of San Antonio could see that they were not sharing in Seligmann's good fortune. The workers labored for hours each day in poorly lit areas with no ventilation, restrooms, or washbowls, a fine, brown dust from the pecans filling the air around them. As a result, San Antonio's tuberculosis rate rose to more than twice the national average. The workers would soon come to realize that without them, Seligmann would fail as well. In the winter of 1837 to 1838, workers were presented with yet another one-cent pay cut. This led to the largest strike in the city's history on January 31, 1938.

Seligmann had little sympathy for his workers. "The Mexicans don't want much money. . . . If they get hungry they can eat pecans. . . . The Mexicans have no business here anyway. They flock into San Antonio with their kind, and then they cause labor troubles or go on relief at the expense of the taxpayer," he remarked at a state hearing in San Antonio on the pecan industry. "The Mexican pecan shellers," said Seligmann, "eat a good many pecans, and five cents a day is enough to support them in addition to what they eat while they work."

Emma Tenayuca, a San Antonio native and pioneering labor activist, initially organized the strike. She was a well-known figure in San Antonio politics, primarily as Communist head of the San Antonio chapter of the Workers Alliance of America. Although she had no official title, Tenayuca, then only 23 years old, was elected by a committee to head the strike as it began.

Prior to the strike, the wages of those workers shelling pecans were dropped by one cent, and those of crackers were reduced from fifty cents to forty cents per 100 pounds. As the threat of the strike grew, San Antonio officials, particularly Chief of Police Owen Kilday, voiced strong opposition to such a disturbance in their city. Initially, all the while he was dispersing demonstrators and arresting picketers, Kilday was claiming there was no strike. Because of mass arrests (more than 700), the strike generated national and international attention. During one week in February 1938, 90 striking workers were arrested. When he did finally acknowledge the strike, Kilday stated that it was part of a "Red plot" to take control of the west side of San Antonio.

After Texas governor James Allred's request for an investigation into possible civil rights violations, the Texas Industrial Commission ruled that the police were unjustified in their interference with the right of peaceful assembly. Finally, in March 1938, the strike came to a peaceful end when both sides agreed to a settlement of seven to eight cents per hour. Congress soon intervened with the passing of the Fair Labor Standards Act of 1938, which established minimum wage at 25 cents per hour. Following the establishment of minimum wage, worker organizations began to fear that the rate of 25 cents an hour would encourage shelling operators to once again invest in mechanization over human labor. They asked Congress for an exemption to the Fair Labor Standards Act but were denied. Within three years, cracking machines had replaced 10,000 workers in the shelling plants of San Antonio.[63]

Today, although Texas has abundant commercial orchards planted to improved varieties, approximately 30 percent of its annual production of pecans remains native seedlings, referring to nuts from trees that grew in place from a pecan that germinated, usually along the bottomland ridges lining the state's rivers and streams. Unlike improved-variety orchards planted in uniform rows, native-seedling orchards are literally carved out of the forest and the trees are scattered at random. Since they do not produce as consistently and do not command the premium prices of most improved varieties, seedling orchards usually do not receive the high inputs of fertilizer, insect and disease control, and irrigation with which improved orchards are managed. Managing a native pecan grove can also be a bit of a gamble because of the frequent flooding that can occur during harvest time, destroying the crop. In a sense, after the native brush is cleared away, most of these trees produce a truly organic crop. The growers simply harvest what nature provides.

Of the 600,000 to one million acres of native pecans in Texas, about 40,000 acres are managed consistently.[64] Each native pecan grove is different depending on its proximity to one of Texas's large rivers. The best orchards tend to be those closer to the rivers, where there is better soil and the trees have greater access to water than those on small tributary streams and creeks.

The first step in bringing a native pecan grove into production is to remove all the competing trees and vegetation from the bottomland, leaving only pecans. Native pecan bottoms are also full of oak, elm, hackberry, and mesquite, which must be cleared away to provide the pecan trees maximum sunlight, soil moisture, and access to nutrients. Aside from other species, poor-producing, weak, diseased, or damaged pecan trees are removed as well so that the remaining trees have the best chance for good production. Just as in an orchard of improved pecan varieties, sunlight is required on at least 50 percent of the orchard floor for optimum production.

Managed native pecan orchards do well to produce 1,000 pounds per acre on average. Often native groves may produce such a crop in one year and have virtually nothing the next. This is about half of what a productive orchard of improved varieties can produce. Occasionally, you will hear tales of large old trees that produce 800 to 1,000 pounds per tree. While this is certainly possible, it is rare, and to be sure, they do not produce such yields every year. For this reason, management practices are more intense in the year that the grove bears a crop. Of course some native seedling trees produce good crops of exceptional quality. These are often singled out and propagated. With the right amount of testing and promotion, these may eventually become improved varieties. Such trees are most definitely the exception rather than the rule. Although fungicides are not usually required, the pecan tree's ancient nemeses, the pecan nut casebearer and the pecan weevil, can be particularly damaging in native bottomland orchards and are often controlled in the most intensively managed groves.

Considering the place that pecans occupy in the heritage of Texas, it is fitting that native pecans account for such a large percentage of the state's production. The autumn bounty found here, in the shaded river bottoms, has always sustained Texas, and it likely always will. The same holds true for the pecan's other native areas, including Louisiana, Mississippi, Oklahoma, Missouri, and Kansas.

Building a Better Pecan

Aside from their value in generating the native pecan crop, the great stands of native pecans harbor an invaluable resource for the future of pecan production in the form of the wide array of genetic material they hold. Though no doubt diminished from its former richness, the vast genetic diversity harbored within these groves and bottomlands of the pecan's native range remains a unique preserve. Likewise, we can find value in the wild pecans sprouted along fence lines and wooded edges in the pecan's extended range. Considering the comparatively narrow diversity found in the majority of other world crops, this bank of material is indeed rare.

In agricultural settings, diversity compensates for myriad problems. Nature is complex. Human agricultural practices tend to take this complexity and simplify it for our own use, packaging it into a form that we can more easily understand and manage. This simplicity, though, has its tradeoffs, the most notable being loss of resilience. Native pecan populations hold diverse forms of trees that have coevolved with pests, diseases, and varying environmental conditions over millennia in the escalating race for survival. For as long as humans have grown pecans, we have selected for characteristics from wild populations of trees that improved their productivity, quality, and ease of management. In many areas where pecans are grown, producers maintain a surprising degree of diversity in their orchards, even after more than 100 years of production. For many other crops, a large portion of the genetic diversity from which they originated has vanished. We are fortunate that the native pecan stands provide an opportunity to continue gleaning diverse traits from wild populations, enabling the development of new varieties or the discovery of individual trees that can be propagated for use in orchards. The preservation of this diversity across the pecan's range becomes increasingly valuable in order to meet the ever-evolving challenges of disease, insects, drought, and cold tolerance, among others.

Without improved varieties, the pecan industry of today would not exist. As propagation techniques like those of Antoine, E. E. Risien, and others were mastered and the early nurserymen of Mississippi began to find superior seedling selections, the pecan industry developed legs. While many of the early varieties like Stuart, Schley, Pabst, Alley, Desirable, Van Deman, San Saba, Success, and others were of enormous benefit to a fledgling industry, only Stuart, Desirable, and, to a lesser extent, Schley are still grown in any significant numbers.

Plant breeding is a long, complicated process that involves selecting male and female parents, collecting pollen, crossing, harvesting and planting the resulting seeds, and evaluating the plants that grow from the seeds and the fruit that grows from the plants that grow from the seeds. This process can take a long time for annual crops, but for a long-lived, nonprecocious perennial plant like a pecan, decades pass, literally, before the fruits of a breeder's labor are proven.

Louis D. Romberg grew up walking back and forth to school through the pecan bottomlands of North and Middle Darrs Creeks in Bell County, Texas. At the time, there was little indication that this schoolboy climbing trees and eating pecans along the creek bank would become the world's first "official" pecan breeder. Romberg graduated from Texas A&M in 1921 and was employed by the Texas Department of Agriculture in 1923, where he worked as an assistant to J. H. Burkett. In 1931 he was hired by the USDA Bureau of Plant Industry and appointed to a position with the USDA's Pecan Research Laboratory at Austin. The program was moved to the US Pecan Field Station in Brownwood in 1938 and eventually came under Romberg's direction.[65]

Romberg developed a technique to speed up the slow work of pecan breeding. By grafting three-month-old hybrid seedlings onto older, bearing trees he was able to generate fruit in 2 to 3 years and evaluated trees in 6 to 12 years. This process dramatically reduced the time required to develop a new cultivar. Without it, the whole process could take up to two decades.[66]

Romberg's greatest contribution was the building of the USDA pecan breeding program at Brownwood, Texas, where he made 10,000 crosses, resulting in the release of 25 new cultivars (to date), beginning with the first pecan cultivar released from the program, Barton, in 1953. Barton resulted from a Moore × Success cross that Romberg made in 1937 in the orchard of John Barton Sr. of Utley, Texas. Initially, all USDA-released pecan cultivars were to be named for individuals who had made significant contributions to the pecan industry. But following the release of Barton, the decision was made to name all USDA pecan cultivars for American Indian tribes, in recognition of the role played by Native Americans in disseminating the pecan. The first of these "Indian varieties," as they are commonly called in the pecan industry, was Comanche, released in 1955.

Many believe that Romberg's original objective was to find a replacement for Burkett, a popular Texas variety at the time. Burkett had an excellent taste but suffered from disease susceptibility and low yields, and it was particularly hard to shell out into complete kernel halves. Romberg made a number of crosses of Burkett with varieties like Schley and Success, seemingly in an attempt to increase the quality and production as compared to Burkett. From these crosses, the cultivars Apache and Comanche were developed. Eventually Romberg's objectives shifted to include precocity and size. He began using precocious cultivars such as Moore, Clark, Curtis, Mahan, Brooks, Major, Starking Hardy Giant, and Evers as parents, along with those that produced large nuts such as Success, Odum, Mahan, Mohawk, and Desirable. [67]

Following Romberg's retirement in 1979, the USDA pecan breeding program was led by George Madden, a native of Weslaco, Texas, along the Mexican border. Madden tested and promoted many of the selections resulting from Romberg's crosses. Under his watchful eye, six of Romberg's cultivars were released. Working under Romberg's wing for a while, Madden was involved in several other USDA releases of which Romberg was the primary author. Madden, along with Harry Amling of Auburn University in Alabama, also developed an improved method of training young pecan trees to avoid weak crotch angles and limb breakage. By training the tree to a central leader, removing the uppermost buds at each node, and then selecting scaffold limbs, growers could help the tree develop a stronger and more resilient structure.

For a few years in the 1990s, the USDA "Indian varieties" developed a bad name in the southeastern pecan region. In the late 1960s and 1970s, the USDA released a number of exciting varieties resulting from some of Romberg's crosses, including Cherokee, Cheyenne, Chickasaw, Shawnee, Shoshoni, and Mohawk. Most of these varieties were extremely precocious and prolific. It was expected that several of them could be planted in high density and maintained through pruning and management at such spacing. Based on the early appearance of many of these varieties, a large number of trees were planted throughout the Southeast. By the late 1980s to early 1990s, most of these plantings had failed to be productive or profitable. The intense scab pressure of the region combined with severe overbearing led to poor quality in an "on" year and virtually no crop the next. However, these traits didn't manifest themselves fully until the trees reached maturity, by which time many people had invested a large amount of money in them. While many of these crosses failed to stand the test of time in the Southeast, they helped stimulate a strong interest in and realization of the value of pecan cultivar diversity. Most were top-worked to other varieties, cut down, or abandoned.

For some, those varieties that show an adequate level of scab susceptibility may still hold some promise. With mechanical hedging, a common practice in the western United States, tree size can be managed more readily. Some pecan producers feel that through hedging and mechanical fruit thinning, tree size and alternate bearing problems can be managed.

By far, the most successful varieties released from the USDA pecan breeding program have been Wichita and Pawnee, both of which originated from crosses made by Romberg. Wichita resulted from a cross between Halbert and Mahan made in 1940. It was released in 1959 and quickly became probably the most widely planted cultivar released from the USDA program for a time. At one point it was grown in virtually every region of the US pecan belt. Large acreages of Wichita were planted in Georgia during the late 1960s and early 1970s planting boom. However, under the humid conditions of the Southeast, it was found to be highly susceptible to pecan scab, which led to its demise in that region. The tree performs much better in the arid climate of the western United States and is still recommended in West Texas, New Mexico, Arizona, California, and southwestern Oklahoma.[68]

To date, the crown jewel of the USDA pecan breeding program has been Pawnee. Resulting from a 1963 cross Romberg made between Mohawk and Starking Hardy Giant, Pawnee was released in 1984. Graftwood of Pawnee was widely disseminated in the early 1970s before it was even released. Known as selection 63–16–125 at the time, Pawnee produced a large, early-maturing nut and producers jumped at the opportunity to have this cultivar in their orchards. As a result, Pawnee is likely now the most widely planted and recommended pecan variety.

But what made Pawnee such a popular choice? The answer is simple: no other large pecan can compete with Pawnee on the early market. Harvested most years in mid to late September in Georgia, Pawnee commands exceptionally high prices because it is the first variety of the year to be harvested. It generates a price benefit of as much as fifty cents to one dollar per pound or more over most other varieties. But Pawnee can't be produced, especially in humid areas, by just any pecan producer. Its high susceptibility to pecan scab and early crop maturity require attention to timely fungicide applications and more rigorous insect scouting. The tree sometimes overloads itself with nuts at maturity, leading to alternate bearing and poor quality if not managed properly. Although pecan scab is a serious problem with Pawnee, under the right conditions, this variety also appears to be able to produce good-quality pecans with more scab on the nut than do other cultivars. This condition likely results from the late development of scab on the nut after the shell has hardened and the kernel is less susceptible to injury. Recommended in almost every pecan-producing state in the nation, Pawnee is now planted from coast to coast along the US pecan belt.

The USDA pecan breeding program continued under the direction of Tommy Thompson until his retirement in 2012. Thompson continued to make crosses and develop new and improved pecan cultivars using twenty-first-century methods. With the advent of new technology, the pecan breeding program began to explore its mission in more ways than just the classical breeding program. In addition to developing and testing new varieties, the program now seeks to conserve and characterize the genetic diversity not only of pecans but of all hickories, a task assigned to Dr. L. J. Grauke.

Established in 1984, the National Clonal Germplasm Repository for Pecans and Hickories grew from pecan varieties collected by Romberg for use in
his breeding program. Grauke has added to this collection by including wild
species and domestic varieties of pecan and other hickories from around the
world. While collections have been made from about 17 species in the *Carya*
genus, the most thorough collection has come from native pecan populations
ranging from Oaxaca, Mexico, to Illinois. Pecan seedling trees grown from
the collected nuts were planted in orchards in Brownwood and College Station, Texas, as well as Byron, Georgia. These orchards are the source of valuable information regarding the influence of geographic origin on the genetic
diversity, phenology, and performance of pecan trees. Grauke and Thompson
have developed and used modern molecular techniques like microsatellite
markers to allow verification of variety identity and to confirm the parentage
of controlled crosses.

According to Grauke, the efforts of the Germplasm Repository have led
to a better understanding of the role played by interspecific hybridization,
which in most cases leads to more vigorous offspring in the pecan breeding
program. By conserving and maintaining the pecan tree's diversity, breeders
can continue the process of improvement. In an ironic twist, Grauke acknowl-
edges that in the development of improved cultivars that may be economically
rewarding, some genetic diversity is lost when well-adapted and uncharacter-
ized native stands are replaced with grafted orchards. As a result, one of the
long-term goals of the Germplasm Repository is to include the designation of
appropriate, regionally diverse reserves on conserved land where the raw materials of diversity can be retained.[69]

While some pecan varieties generated by the USDA program have proven
unsuitable for southeastern growing conditions, there have been a number of
releases since the late 1960s that hold promise in this region. These include
Pawnee, Oconee, Caddo, Creek, Kiowa, and Kanza. In fact, Caddo was developed from a cross between Brooks and Alley made in 1922 or 1923 at the USDA
Pecan Field Station in Philema, Georgia, near Albany. It was released by
Romberg in 1968.[70] To date, the USDA pecan breeding program has resulted

in the release of 28 pecan varieties generated from controlled crosses. This work has been invaluable to the development of the US pecan industry, and the program continues to develop varieties to help move the industry forward.

Because many of the varieties generated by the USDA program appeared more suited to arid western climates, the University of Georgia hired Dr. Patrick Conner to begin development of a pecan-breeding program. The goal of Conner's program is to develop profitable pecan varieties adapted to the humid conditions of Georgia and the southeastern United States. He selects his crosses based on large nut size, shelling characteristics, kernel color and quality, cluster size, and insect and disease resistance. Dr. Darrell Sparks, long-time pecan researcher at the University of Georgia, developed and released several cultivars around the time of his retirement in 2009, these resulting from crosses he had made in the early 1980s. The first of these to be released included Byrd, Morrill, and Cunard.

Fourteen years and several thousand crosses later, Conner's program is nearing the point at which a number of promising selections are ready for testing in the orchards of various cooperators. Pecan breeders such as these continue to work quietly and patiently behind the scenes. Remember, pecan breeding is a painfully slow process. In general, the process goes as follows: Pecan breeders want the female flowers of their chosen parent to be pollinated only by the male parent of their choosing. In order to accomplish this feat, the female flowers are enclosed in white paper bags or sausage casings before they become receptive to pollen. A plug of cotton is normally maintained at the base of the bag. Later, when the female flowers are receptive, a hypodermic needle is inserted through the cotton into the bag and pollen is puffed inside where it settles, hopefully on a receptive female flower. Of course, before this can be accomplished, pollen is collected from mature catkins and stored in a refrigerator until ready for use.

The nuts resulting from these controlled crosses, if they survive, are then planted into a field at close spacing for one to two years, when they are dug up and transplanted to another field where they will be grown until they fruit. They are then evaluated for a number of years based on the criteria set forth by the breeder. If they make the cut, they are selected for evaluation on a larger scale, possibly in different locations under various growing conditions. After a few years of these evaluations, if they make the next cut, the selections are usually released and awarded full cultivar status.

This painstaking process is necessary to increase the chances of a variety standing the test of time. Like all things, a variety's status may change through the years. Diseases and insects adapt to overcome the tree's defenses. In addition, a mature tree behaves very differently from a young tree, and poor traits will often manifest themselves only with enough time and with more widespread dissemination. The more closely pecan breeders can monitor and screen for the traits they seek, the better the cultivar they will produce. The wonderful genetic diversity found in the pecan tree itself holds the secret that pecan breeders attempt to release.

A Tree Grows in the Desert

Development of the pecan industry on a large commercial scale in the Desert Southwest lagged somewhat behind that of the rest of the pecan belt east of I-35. As mentioned earlier, irrigation was the key. The water-loving pecan tree struggles to survive in this arid region without it. The first pecan trees grown in New Mexico were likely seedling trees that resulted from pecans brought to the area from central Texas and north-central Mexico in the late 1800s and early 1900s.[71]

The oldest planting of improved pecan varieties in New Mexico is found on land owned by New Mexico State University in Mesilla Park. These trees were planted by the first director of the University's Agricultural Experiment Station, Fabian Garcia.

Garcia was born poor in Chihuahua, Mexico, in 1871. At the age of two, he was orphaned and moved to New Mexico under the care of his grandmother, Doña Jacoba, who worked as a housekeeper in the towns of Lorenzo and Georgetown before settling near Old Mesilla. Here, she took a position in the home of Thomas Casad, an event that would change young Fabian's

FIGURE 8 Fabian Garcia, father of the New Mexican food industry, credited with paving the way for the state's chili, cotton, onion, and pecan industries. Used with permission of the New Mexico State University Library, Archives and Special Collections.

life and that of the Mesilla Valley as well. The Casads were a prominent pioneer family and owned the 5,000-acre Santo Tomás Spanish Land Grant. The Casads took a liking to the young Garcia and provided him with private tutors and later sent him to Las Cruces College, which would eventually become the New Mexico College of Agricultural and Mechanical Arts and later New Mexico State University. Garcia was a member of the first graduating class of the New Mexico College of Agricultural and Mechanical Arts in 1894. From there, he went on to study at Cornell University in New York before returning to the Mesilla Valley, where he served as an assistant professor at his alma mater and earned a master's degree in horticulture. In 1904, he became a professor of horticulture.

Two formative events occurred in the life of Fabian Garcia in 1907. He was married to Julieta Amador, from one of the area's most prominent families, and he began work to hybridize a more standard chili pod. Sadly, Julieta would pass away after only 13 years of marriage, and Garcia would never re-marry. He instead dedicated his life to his work.

The early cultivated chilies of New Mexico were small and were used primarily for spices. Eventually larger chilies were grown, but there was no uniformity in size or shape, which made it difficult for farmers to determine which variety they were growing. Garcia's work, culminating in the release of the first modern chili variety, Number 9, in 1921, revolutionized the chili pepper industry and became the foundation for subsequent varieties.[72] Garcia also experimented with cotton, sugar beets, and onions, and in 1913 he planted 35 varieties of pecan trees to test in the Mesilla Valley, some of which remain today. This four-acre orchard holds the original improved varieties planted in the area.

Fabian Garcia became the first director of the state Agricultural Experiment Station in 1914. When the land for the present horticulture farm on which the pecan trees were planted was purchased, it was Garcia who personally signed the note for the loan. He paid it off by growing watermelons, watering them with hand-operated "pitcher pumps." Garcia's enthusiasm, productivity, vision, and kindness endeared him to New Mexico's farmers and made him a legend among those he served. Today Garcia is known as the father of the New Mexican food industry and is credited with paving the way for the state's chili, cotton, onion, and pecan industries. Garcia's plaque in the American Society for Horticultural Science Hall of Fame reads: "Dr. Fabian Garcia, a man of humble origins, but a gentleman of extraordinary achievements."

Today more pecans are produced in Doña Ana County, New Mexico, around Las Cruces than in any other county in the United States. The greening that ultimately made pecan production possible here resulted from a series of engineering feats that began about 1905. Two large storage dams, a cluster of small diversion dams, two flood-control dams, nearly 600 miles of irrigation canals, and over 450 miles of drainage channels brought agriculture to this region. Much of the irrigation water comes from Elephant Butte Reservoir, a 36,000-acre lake on the Rio Grande near Truth or Consequences, New

Mexico. The reservoir is impounded by Elephant Butte Dam, the largest dam ever constructed at the time of its completion in 1916, and captures water draining from nearly 29,000 square miles of snowmelt in southern Colorado and northern New Mexico. It is a major component of the water system, delivering a predictable water supply to southern New Mexico, West Texas, and Mexico.

Although Fabian Garcia's orchard sparked some interest in small orchards around Mesilla, commercial production of pecans at the magnitude we know today was nonexistent until the 1960s. The burgeoning pecan industry in the Desert Southwest today was simply waiting for someone with the means and imagination to start the ball rolling. That man was Deane F. Stahmann.

If a traveler leaves El Paso, Texas, driving the Trans-Mountain Road, he crosses over into New Mexico and eventually reaches a dead end on Highway 20 at the Rio Grande. If he heads north and takes a quick left just past the small town of Canutillo, he will find himself in New Mexico along historic Highway 28 (also known by its more elegant name, the Don Juan de Oñate Trail), which will lead him through the fertile Mesilla Valley. Along the way the traveler will pass dozens of farms, almost every one with at least a few acres of pecan trees. After a few miles he will find himself driving through the center of a large pecan orchard, and just before crossing back over the Rio Grande, he will see Stahmann's Country Store on his left. At one time, the traveler would have been in the center of the largest pecan farm in the nation, Stahmann Farms.

Stahmann Farms has its unlikely origin in central Wisconsin near the town of Bruce. Here a buggy maker and beekeeper named W. J. Stahmann, along with his wife and two children, struck out from the Midwest in 1909, traveling on a barge down the Mississippi. Like many people, the Stahmanns moved west because of the effects of the arid desert climate on tuberculosis. Stahmann's wife, Hannah, had recently been diagnosed with the illness. Hearing of the potential recovery from tuberculosis offered by dry climates, W. J. Stahmann chose to uproot his family in the hope of saving his wife's life.

Stahmann sold the barges in Arkansas and the family made its way across Texas, settling in Fabens, near El Paso. Here, Stahmann began farming cotton, tomatoes, onions, alfalfa, and rabbits. His cotton and tomato operations grew rapidly, which led to a canning plant and four cotton gins. Stahmann also set up the beehives he had brought with him from Wisconsin and continued making honey in his new desert home.

Worried that the brackish and low-quality irrigation water on their Fabens farm would become too saline for crop production, Stahmann moved his family north to the Mesilla Valley of New Mexico. In 1926, W. J. Stahmann purchased 2,900 acres in southern New Mexico. Along with his son Deane, Stahmann cleared the land with mule teams and began planting cotton. This new location pleased Stahmann, primarily because of its close proximity to Elephant Butte Dam. The Stahmanns' land purchases made up most of the Santo Tomás Land Grant, once owned by the Casad family, who had taken the young Fabian Garcia under their wing.

When W. J. Stahmann died in 1929, Deane sold several parcels of land, keeping about 2,900 acres. He later acquired the 1,100-acre Snow Farm, which is now within the city limits of Las Cruces, just north of the Stahmann headquarters. As the Depression arrived, the price of cotton fell from 23 cents per pound to 6.5 cents per pound within a decade. Much of the Stahmanns' land remained sand dunes and mesquite bushes.

Deane Stahmann was first and foremost a cotton farmer. The Stahmanns' pecan empire grew from a simple whim. In 1932, Deane ran into a man selling pecan trees that the original buyer in El Paso could not afford to buy. Stahmann bought 2,000 trees and planted them at the south end of the farm. But these pecans, of course, brought no income for the first five to seven years. So, initially, the trees took a backseat to the farm's other crops: cotton, alfalfa, lettuce, cantaloupes, cucumbers, and cattle, which grazed the orchard.

For a while, cattle would continue to roam the orchards. But when Deane realized that the cows were compacting the soil and causing drainage problems, the cows were removed. Deane had thought the natural fertilizer and weed control the cows had contributed were valuable. In their absence, he looked around for another solution. He settled on the introduction of 25,000 white Chinese geese in 1953, which provided the same result without the soil compaction created by the heavy cattle.

Stahmann's flock of geese increased in size from 25,000 to several hundred thousand once he noticed that the birds could weed the cotton without damaging the plants. Other farmers in the area took notice and soon Stahmann was leasing out geese for weed control. As preplant herbicides were later developed, the geese were no longer needed in such numbers and Stahmann discontinued their use. At one time, Stahmann Farms was the largest producer of geese in the world and some of the birds were processed and marketed. Today the geese and their service are memorialized in a mural adorning Stahmann's Country Store, depicting the birds at work in the pecan orchard.

Deane Stahmann's innovative mind continued to work on the development of improved cotton varieties. One of these, Del Cerro, became a major crop in South Africa. He would establish research farms in Mexico and Jamaica in order to study his cotton varieties year-round.

In the 1940s, the Stahmanns built an on-site pecan processing operation, and by the 1950s two pecan shelling plants were handling 8,000 pounds of pecans a day. Cotton remained the primary crop on the farm until the late 1960s. As the pecan orchards reached full production, cotton was pushed out, and by 1968, pecans had replaced cotton as the main crop on Stahmann Farms. As many as 1,000 workers were employed during the harvest season to bring the crop in by hand, and Stahmann Farms took on the character of a small town, with a central office, slaughtering plants, blacksmith shop, machine shop, nurse's clinic, and housing; the commissary was housed in the building that now serves as Stahmann's Country Store.

Deane Stahmann's agricultural innovations were not exhausted solely on cotton. One issue facing pecan farming was the relatively low production per acre of land compared with other western crops. Because pecan trees grow to such great size, a single mature tree requires ample room, so pecan producers removed trees when they became crowded. Stahmann sought to solve this problem by controlling the size of the trees and retaining trees in the orchard at a higher density. He did this by pioneering a technique called mechanical hedging.

In the 1960s, Stahmann began experimenting with this process by interplanting trees in an older orchard, yielding a density of 48 trees per acre.

During the initial stage, he pruned the larger trees back each year using a variety of methods including tree topping and heavy pruning until they were all about the same size. Although the current practice is somewhat modified from Stahmann's original experimental pruning methods in terms of degree and design, mechanical hedging is now considered a standard management practice throughout the western United States.[73]

As time passed, the Stahmann orchards grew to include some 128,000 Western Schley and Bradley pecan trees. As more-modern processing facilities replaced the first two and mechanical shakers were developed, the Stahmann's work force was reduced. Eventually, the Stahmanns shut down their processing facilities. Until the late 1980s, Stahmann Farms was the largest contiguous pecan orchard in the world, a title now held by Farmers Investment Company, or FICO, in Arizona. However, Stahmann Farms remains a highly productive pecan farming operation with national and international interests. Without the vision and innovation of Deane Stahmann, the western US pecan industry would not be as significant as it is today.

New Mexico produces an average of 60 to 70 million pounds of pecans a year on about 40,000 acres. This places it second or third in the country in production (depending on the year) and fourth in terms of pecan acreage. This high production from a smaller pecan acreage than some of the other pecan-producing states is, at least partially, a result of the high-density planting and mechanical hedging developed by Deane Stahmann. A scholarly man who based his farming and business theories on extensive reading, Deane Stahmann believed in efficiency and was willing to experiment, which played major roles in the farm's prosperity. Even so, the dedication of its owner was the most valuable component of the success of Stahmann Farms. This is best summed up in one of Stahmann's favorite sayings: "The best fertilizer a pecan tree can have is the shadow of its owner."[74]

Like Stahmann Farms, Farmers Investment Company has a rich heritage of pioneering pecan production in the Desert Southwest. Keith Walden, the oldest of five children, was born to a pioneer family on July 4, 1913, in Santa Paula, California. The patriarch of this family, Arthur Frisbie Walden, was a banker who also owned a citrus farm and partnered with a brother on a San Fernando Valley fruit and vegetable farm. Here young Keith learned the merits of hard work and plunged his hands into the soil, developing a lifelong love and respect for the land. At the age of 16, Walden was stricken with polio, forcing him to spend a year in bed. The illness gave Walden time to reflect on what he wanted to do with his life, and he determined that upon recovery, his would be an active life of association with the land. In his view, land was a resource best used to grow food and fiber for people. He recognized three basic human needs—food, clothing, and shelter. Around these needs, he would build a life, he determined. Walden eventually recovered from the disease and at the age of 24, he developed a 10-acre citrus nursery and went on to work as a plant pathologist for the Limoneira Ranch. Later, a stint managing the Ford-Craig Ranch in San Fernando during World War II allowed him to save enough money to purchase 960 acres in the Tulare Lake bed.[75]

By the mid-1940s, Walden's own diversified farming operation included cereal grains, oilseed crops, cotton, sugar beets, vegetables, melons, potatoes, and alfalfa. He incorporated this large operation as Farmers Investment Company, better known today as FICO, but land values in California were skyrocketing and Walden began to look for cheaper land to farm elsewhere. He found it in Arizona. The 10,000-acre Continental Farm, lying in the Santa Cruz Valley 25 miles south of Tucson, was founded in 1915 by Bernard Baruch, Joseph Kennedy, and J. P. Morgan. Their idea was to grow *guayule* for rubber, out of fears that Germany would try to block rubber imports by cutting off shipping lanes.

Of 2,000 species of rubber-producing plants, today only two are grown commercially—*Hevea brasiliensis* and guayule. While the two species have had parallel histories of development as crops, *Hevea* has been the primary focus of the Rubber Research Institute of Malaya, which has generated higher yields and more reliable production for *Hevea*, a native Amazonian plant. Guayule's production, on the other hand, has suffered from weak and intermittent research efforts through the years. *Hevea* is the world's dominant rubber crop today. However, the three business titans chose guayule for the Continental

Farm because the plant is native to the limestone bajadas and hillsides of the Chihuahuan Desert and is well suited to the conditions in southern Arizona. While the Continental Farm was successful in producing an average of over 1,400 tons of rubber per year, the project was abandoned by the end of World War I.

The Continental Farm was sold in 1922 to Queen Wilhelmina of the Netherlands, who leased the land out for the production of cotton. After FICO's purchase of the Continental Farm in 1948, the initial crops grown on the farm were cotton, alfalfa, silage, wheat, barley, lettuce, and watermelons. Cattle were introduced in 1953, and FICO's cattle operation grew to 20,000 head before it closed its feedlot in 1976.

Sometime in the 1960s Keith Walden became concerned about the rumblings circulating over the development of synthetic fibers by companies such as DuPont and Union Carbide. Afraid that the new materials would lead to a depressed cotton market, Walden began experimenting with various alternative crops such as walnuts, almonds, pistachios, nectarines, apricots, pecans, and grapes. Of these, pecan and grape production rose to the top under Walden's management style. Because they had a longer window of harvest and could be mechanically harvested, Walden decided that pecans were best suited to FICO's operation and in 1965, the company's cotton farms in Sahuarita and Continental began their transformation into the world's largest irrigated pecan orchard, with some 6,000 acres of orchard land containing 106,000 trees.

Today FICO's orchard holdings include the operations in Arizona as well as about 1,000 acres in Georgia known as the Blue Three or "Blew Three" orchard at Pecan City near Albany. Legend has it that the farm received its unusual name when Walden made the comment that he had "blown 3 million" dollars when purchasing the orchard. Today, it is considered one of the finest producing orchards in the state. Pecans are now grown in 11 of Arizona's 15 counties on a total of more than 19,000 acres, mostly in the southern half of the state. The state produces around 20 million pounds per year and ranks fourth in national production, thanks largely to FICO.

FICO, like Stahmann Farms, has been instrumental in the development of mechanical hedging as a standard production practice for western-grown pecans. Since 1975 the company has continually experimented to refine its methods of hedging, topping, selective pruning, and tree removal to maximize production, quality, and sunlight. Conservation has long been a priority for the Walden family and FICO. This is most readily observed in their efforts to enhance the water-use efficiency of pecans in the area because of the limited water availability in the dry desert climate. In 1980, FICO began using laser-leveled technology in its Arizona orchards to maximize irrigation efficiency. This technique led to a 20 percent reduction in water use in the FICO orchards. FICO's commitment to community and conservation has given the Walden family and its company a deserved reputation as excellent stewards and vital partners in their surrounding communities. In addition to being the largest integrated grower and processor of pecans in the world, FICO is also the largest producer of organic pecans on the planet.[76]

The greatest challenge facing western US pecan production, as one might expect, is water. In 2013, drought conditions led to some of the lowest levels of available water in years. Drought has taken its toll on agriculture in southern New Mexico and West Texas repeatedly throughout history. In the late 1940s, drought set in, but the suffering was buffered by the return of rain in the mid-1950s. At the turn of the twenty-first century, drought began reducing reservoir levels again, prompting lawsuits between Texas and New Mexico over the water historically allocated to each state by the US Bureau of Reclamation.

Agricultural water users like pecan farmers are allocated only a certain amount of water from irrigation projects each season to irrigate their crops. The 2013 allocations were the lowest in history. Spring estimates of runoff for April to July of that year were 22,000 acre-feet, about 5 percent of the historical average. El Paso Valley farmers were informed they would be allowed only six inches of irrigation water for the 2013 crop. Mesilla Valley farmers would receive only three and a half inches, a paltry amount compared to the four or more acre-feet they had received in the past. Fortunately, most pecan farmers could supplement their water allotments with well water, but this created problems with salinity for some. Water quality in the Mesilla Valley is good but farmers must dig as deep as 500 to 600 feet to reach it, an expensive investment that adds another 10 to 15 percent to their expenses. El Paso Valley producers, however, must pump water from wells with high levels of sodium, a scenario that, according to some, is unsustainable for extended periods.

The drought magnifies the salinity problem because as more groundwater is pumped, more water is pulled from deeper levels where the water may be more salty. Water is also sucked from the edges of the aquifer, where salts are heavily concentrated. Therefore, the more water is pumped, the more salty it becomes. The West's water woes are far from behind it and remain the most significant obstacle to the further expansion of its pecan acreage and agriculture in general.

The Gulf Coast

Although much work, sweat, and money are involved, something elegant remains about growing pecan trees. It's in the sound of the leaves in the wind, the texture of the bark, and the coolness beneath the canopy. Perhaps no other place embodies this aesthetic as does the birthplace of the Alabama pecan industry, Baldwin County, a place where agriculture meets the sea.

Located in south Alabama along the Gulf of Mexico and Bon Secour Bay, Baldwin County was home to early European exploration and settlement in the southern United States. From the mid-sixteenth century until the American Revolution, Baldwin County land was transferred, not always peacefully, from Native Americans to Spain, France, Britain, and finally, to the United States. At the dawn of the nineteenth century, the area was the center of growing hostility between local Creek Indians and settlers and is historically marked by a massacre at Fort Mims in August 1813, in which more than 250 settlers and Indians died. The area's economy at the time was founded on agriculture and deerskin trading among the Creek, settlers, and Europeans. Mobile was home to export houses that shipped the deerskins across the Atlantic to Europe. Settlers found land cheap and fertile for the production of indigo, tobacco, and rice. Cattle production became a profitable business, and the dense forest stands made timber, turpentine, potash, and tar valuable exports.

Baldwin County attracted many Italian, German, Greek, and Eastern European immigrants at the turn of the twentieth century, and in November 1894, Iowa journalist Ernest Berry Gaston and 28 settlers founded the utopian community of Fairhope on the cliff above Mobile Bay. The area's rich history extends even into the local government. The town of Daphne served as the original county seat of Baldwin County in 1787, while it was still a part of the Mississippi Territory. A rival community, Bay Minette, stole the courthouse records in 1900. Since that time, Bay Minette has served as the county seat.[77]

During these many layers of history, the Alabama pecan industry was born. Commercial pecan culture began in 1916, when Scott Higbee and his son Evan planted the first grafted pecan trees reported in Alabama at their farm in Fairhope. This planting consisted of 28 pecan trees of 14 varieties, including Success, Mobile, Columbia, Frotscher, Moneymaker, and a Stuart reported to have a higher kernel percentage and a thinner shell than had been previously observed. Evan would plant 76 more acres in 1921. Forty-three years later, Evan's son Dick would plant another 102 acres and become one of the first growers in Alabama to install drip irrigation. He continued farming pecans into the twenty-first century in spite of outrageous land prices brought on by the booming population and tourist industry drawn to the Alabama Gulf Coast.

Not all pecan orchard plantings have been profitable. From 1954 to 1960, Alton Sherman planted 420 acres of pecans near Fairhope, a planting that was, at the time, the largest pecan orchard in Alabama. Original varieties included Stuart, Elliot, Cape Fear, and Desirable, among others. In the early 1970s, Cheyenne was interplanted in the craze to establish the highly productive and precocious USDA varieties of the time. However, this action proved to be a mistake along the scab-prone Gulf Coast. The owners reported that the orchard never made a profit, forcing them to grow Irish potatoes between the tree rows to keep the operation afloat. It was later purchased by a group of investors and converted to a horse farm.

Pecan production has, however, been a significant industry in Alabama since the 1940s. At one time, pecans were produced in about 25 counties in the southern half of the state. Yet Baldwin and Mobile Counties have always accounted for more than half the state's pecan production. The industry here was built largely on the Stuart, Schley, and Success varieties. During the 1960s and 1970s, growers started planting varieties such as Desirable, Cape Fear, and Elliot. Beginning in the mid to late 1990s, there was a shift to more disease- and pest-resistant varieties like Gafford, Amling, Excel, McMillan, Caddo, Forkert, and Surprize, all of which can be grown with less intensive management.

The Gulf Coast states of Alabama, Mississippi, and Louisiana face constant threat from hurricanes. Since 1871, Louisiana alone has been hit by hurricanes an average of once every three years. Three of the most devastating hurricanes in history, Camille in 1969, Frederic in 1979, and Katrina in 2005, wreaked havoc on the Gulf Coast pecan industry. Hurricane season in the United States extends from June through November, peaking in both number and severity from mid-August through September, a time in which pecan trees are laden with heavy fruit, making the trees and crop particularly vulnerable to damage. Hurricane winds in the southeastern United States initially blow from the northeast, shifting to the southwest with the counterclockwise rotation centered on the eye of the storm. This causes the most destructive winds to occur northeast of the storm's eye. Strong winds coupled with an abrupt shift in wind direction during a hurricane can break limbs, strip leaves and fruit, uproot trees, or twist them, damaging the root system.

Camille, Frederic, Opal, Ivan, and Katrina all took their toll on the Gulf Coast states, chipping away at their pecan industries. Hurricane Camille alone struck Mississippi with such force that its pecan industry was nearly eliminated by that single storm. The year following the storm, Mississippi's pecan production was only about 27 percent of the previous five-year average. It has yet to fully recover to its pre-Camille production levels, with only about two million pounds harvested on 14,000 to 16,000 acres in 2011. Mississippi's tung oil industry, already struggling to compete with cheaper product from Argentina, was completely wiped out by Camille as well.[78]

While Katrina's legacy will likely forever be the loss of life and images of the miserable crowds filling the Louisiana Superdome in the days after the storm struck New Orleans, the pecan industry in Louisiana and Mississippi was hammered by the category 3 storm as well. Fortunately, when the storm struck on August 29, 2005, Louisiana was suffering through the second of two consecutive poor pecan crops. Louisiana's pecan industry consists primarily of native trees, with only a handful of orchards of improved cultivars. The 2004 crop was blistered by pecan scab from early rains in May and June. Since the state's industry is composed primarily of farms with relatively small acreages, the cost of sprayers and chemicals to control the disease on much of Louisiana's pecan acreage is difficult to justify. The stress brought on by disease on these trees likely limited the 2005 crop as well, which turned out to be a blessing in disguise. The moderate crop allowed the trees to take the storm's wind much better than they would have with a heavy crop load weighing down the limbs. The trees came back the following year with a much better pecan crop than was expected following a damaging hurricane year. Louisiana's 21-million-pound pecan crop in 2006 made it the fourth-largest pecan-producing state in the nation that year.

Perhaps no other state's pecan production has been hit so hard by hurricanes over the years as has that of Alabama. The most significant storms to hit Alabama's pecan industry have been Frederic, Opal, and Ivan. In 1979, Alabama's pecan crop was forecasted at 16 million pounds, but on Wednesday, September 12, about two-thirds of that crop was destroyed as Hurricane Frederic blew across Baldwin and Mobile Counties, leveling much of that area's 12,000 acres of pecan. Approximately 60 percent of bearing trees in these two counties alone were lost. Some growers lost as many as 90 percent of their bearing trees. The remaining 40 percent of pecan trees in Baldwin and Mobile Counties were in various states of damage, requiring severe pruning and/or resetting.[79]

Sixteen years later, in October 1995, Hurricane Opal slammed into the Alabama Gulf Coast with wind speeds of over 200 miles per hour, destroying 50 to 80 percent of pecan trees in some areas of the state. Damaging hurricane-force winds returned yet again to the Alabama coast in 2004, taking out 15 to 20 percent of the state's trees.[80] Despite repeated abuse from Mother

Nature, Alabama's pecan producers continue to persevere. The industry is smaller than it once was, but the state's remaining pecan producers press on.

While there is little good that can come from such tragedies as the Gulf Coast states have suffered in repeated hurricane bombardment, it has provided the determined few an opportunity to replace large, old, and difficult to manage trees with more productive, disease resistant, and vigorous trees. Government disaster assistance in the aftermath of Hurricane Ivan aided some growers in their decision to press on. Promising new varieties have reinvigorated what remained of the industry following Ivan. Still, Alabama's pecan industry has yet to recover to its pre-Opal strength. From a record harvest of 61 million pounds of pecans in 1963, Alabama's pecan production has declined, perhaps never again to reach that level. Alabama remains a vital component of the southeastern US pecan industry, however; production in 2011 reached 10 million pounds, making Alabama the number five pecan-producing state in the nation. Although drastically reduced from their former prominence, each of the main Gulf Coast states continues to harbor significant pecan industries and will always hold a treasured spot in the history of this crop.

THE GREAT PLAINS

Like Texas, Oklahoma is home to native pecans, where they have been a staple for ages. While many Oklahoma pecan producers grow improved varieties like Kanza, Pawnee, Kiowa, and Caddo, the backbone of Oklahoma's pecan industry remains its native trees, which account for 90 percent of the state's production. Many orchards in Oklahoma are double-cropped with forage used for hay or for grazing beef cattle. While the volume of pecan production from Oklahoma has taken a back seat to that of Georgia, Texas, and New Mexico, Oklahoma's pecan producers are very efficient, thanks in large part to two notable factors: Oklahoma State University's pecan research and extension program and the work of the Noble Foundation.

Much of the basic understanding we now have regarding the use of fertilizers on pecan trees has come from the research program of Dr. Mike Smith at Oklahoma State University. Smith also developed the technique of mechanical fruit thinning of pecan trees to help alleviate the problems of alternate or irregular bearing and conducted many studies on the basic biology of the pecan tree to help enhance the efficiency of production.

The Samuel Roberts Noble Foundation sprang from the desire of a successful Oklahoma oilman, Lloyd Noble, to give back to the area he called home. Noble had witnessed the effects of the 1930s Dust Bowl, which blew the soil from the plains states away in great suffocating clouds of dust, devastating the once fertile grasslands. This problem resulted from poor farming practices like the failure to rotate crops and control erosion, leading to a mass exodus of farmers and their families from the Oklahoma plains.

With his mother's help in cosigning a loan, Noble borrowed $15,000 for the purchase of a drilling rig in 1921. He quickly developed a reputation as one of the most successful and respected drilling contractors in the United States. While flying from Oklahoma to drilling sites across the United States, Noble could view the eroded gulleys and dying land across the rolling prairie. Lloyd Noble developed a love and respect for the land and its people through his life experiences. A man of uncommon long-term vision, Noble once stated, "We believe that while at times we have felt the overshadowing presence of oil, we are living in an area that is essentially agricultural. This is easily realized when one takes the time to remember that the land must continue to provide for our food, clothing, and shelter, long after the oil is gone."

It is this vision that led Lloyd Noble to establish the Samuel Roberts Noble Foundation in 1945 "for the benefit of mankind." Initially, the foundation established a series of agricultural contests open to residents of Carter and Love Counties in Oklahoma in 1947, which proved to be an excellent educational tool. Eventually, various divisions were established to address critical problems in agriculture from various angles. Today, the foundation has over 360 employees working in three divisions: Agriculture, Plant Biology, and Forage Improvement. The Agricultural Division serves a 47-county area of Oklahoma and Texas with consultation, education, research, and demonstration free of charge. The foundation also has six research farms consisting of 12,000 acres total.[81]

An important portion of the Agricultural Division serves the region's pecan producers, working closely with Oklahoma State University and Texas A&M. Dr. Charles Rohla, a former student of Mike Smith, leads the Noble Foundation's pecan work today, helping to enhance the region's production. The primary challenges to pecan production in this region are ice storms, late spring freezes, drought, and water quality, as well as various insect and disease problems.

Equipment manufacture is another major Oklahoma contribution to pecan production. In 1965 Basil Savage built a rotating off-balance prototype pecan trunk shaker in Madill, Oklahoma. By 1969, he had built and sold 15 shakers. The idea of building a machine to shake pecan trees was not new, but Savage's version was affordable for smaller producers. It was designed to run off the power of a 35-horsepower tractor and effectively solved the problem of waiting until late in the year to harvest pecans until the winter winds simply blew the nuts from the tree. This equipment made smaller-scale pecan production more profitable. Initially, Savage leased the patent and marketed the shaker through Bowie Industries until 1977, when he purchased the Nut Hustler Company with Jim Goforth. This was the beginning of a business that found its niche in helping smaller pecan producers become more efficient by producing affordable equipment for them. A born tinkerer, Basil Savage went on to develop harvesters, sprayers, crackers, and processing equipment. Although still headquartered in Madill, Oklahoma, Savage Industries, once known as Savage Equipment, has developed a worldwide reputation for its products.

Largely through the efforts of Oklahoma State University, the Noble Foundation, and Savage Industries, Oklahoma's pecan producers generate an average of about 15 million pounds of pecans a year and contribute significantly to the nation's pecan production, particularly in the form of the valuable native pecans desired by so many.

While over 70 percent of pecans produced in the United States come from Georgia, Texas, and New Mexico, a valuable and somewhat varied pecan industry exists in what is considered the "Northern Pecan Region." Natural stands of seedling pecan trees exist above the Mason-Dixon Line in Kansas, Missouri, Iowa, Illinois, and Indiana along the floodplains of the Mississippi and its major tributaries. Out of the natural stands of the forested pecan bottoms, people have carved orchards of natives just as they have in Texas, Oklahoma, and Louisiana. However, these pecan trees differ from their southern cousins in that they are adapted to shorter growing seasons and more extreme winter temperatures. Temperatures in the central plains states can fluctuate from -22° to 100°F or more in the summer. In these areas, the frost-free days of the growing season range from 155 to 200, testing the limits of the pecan's cold tolerance.

In addition to native trees, northern pecan producers also cultivate short-season, cold-hardy pecan varieties, many of them originating from wild selections. The short season and cold temperatures require a specific set of characteristics that limit the variety selection that northern producers can utilize. While northern pecan trees have a reputation for cold hardiness and early fruit maturity, these two traits are not necessarily linked. Two of the earliest-ripening pecan varieties, Osage and Pawnee, for example, can suffer from shoot dieback during extreme winters. At the other end of the spectrum, Stuart is capable of tolerating bitterly cold temperatures but requires a longer growing season to mature the fruit.[82]

Over the years, northern pecan varieties have developed a reputation for small, hard-shelled nuts from slow-growing, nonprecocious trees. However, this is a highly limited view of the broad pecan diversity found in the central plains. The native forests here are, in fact, a treasure trove of diversity. Observant northern pecan producers have had ample opportunity to evaluate a large number of wild pecan trees and select the most productive. Northern pecan varieties include Major, Chetopa, Faith, Jayhawk, Colby, Norton, Lucas, Peruque, Giles, Goosepond, and Hirschi.[83]

The Major cultivar became a standard for the northern growing region in the mid-twentieth century and has an intriguing story. Discovered as a chance seedling growing on the farm of Laurie B. Major in Henderson County, Kentucky, the original Major tree was one of thousands of trees growing in a huge natural pecan grove near the confluence of the Green and Ohio Rivers called the Point. Well into the 1950s, carloads of people would descend on the Point each fall to gather pecans from the river bottom on shares. Sometime prior to 1910, Mrs. Major gave J. Ford Wilkinson of Rockport, Indiana, permission to collect wood from one of her trees for grafting onto other trees. Wilkinson climbed the 135-foot-tall tree to reach the lowest limbs 63 feet above the ground. He would later found Indiana Nut Nurseries based on grafts made from this tree.

At some point, a photographer asked Wilkinson if he could photograph him scaling the tree, to which Wilkinson agreed. For unknown reasons, this infuriated Mrs. Major, who promptly swore out a warrant for Wilkinson's arrest, which she carried with her at all times. On his last visit to the tree, Wilkinson was 80 feet high among the tree's branches when Mrs. Major appeared at the base of the tree with a double-barreled shotgun, effectively "treeing" Wilkinson, who scampered to keep the tree's trunk between himself and the gun-wielding widow. When Mrs. Major returned to the house, Wilkinson hurried down the tree, gathered up his graftwood, and left in haste, never to return.[84] A number of varieties released from the USDA breeding program are grown in the central plains, some of which have Major as a parent, including Kanza and Lakota. Pawnee is now a popular cultivar in this region and remains one of the largest nuts that can be consistently produced in the short growing season of the central plains. The short growing season lends a certain advantage to pecan production on the central plains, as the crop escapes many pests of this region. Yet the early-ripening nuts may lead to more problems from some pests, such as the pecan weevil. The pecan weevil, resembling a boll weevil with a much longer snout, bores into the pecan and lays one or more eggs inside. The earlier that pecans ripen, the sooner the weevil's egg laying begins.

Good kernel quality is also a benefit of growing pecans in cool climates. Lower temperatures during the ripening stage tend to limit darkening of the kernel and spoilage of the meat itself, which can be a problem in other pecan regions during an Indian, or late, summer. The cold winters also help preserve the nuts for a longer period, allowing harvest to continue sometimes into

March without a significant loss in kernel quality. These favorable ripening and harvest conditions usually lead to a bright, attractive kernel with good quality.[85] While pecan producers in the northern producing region tend to maintain smaller acreages than their southern counterparts, they also work with nature to become good pecan farmers.

Pecan production as we now know it grew, literally, out of the land from which the crop originated. The progress of development of the pecan from a forest tree to a thriving crop has followed the same trajectory as that of our nation over the course of just over 200 years. Both have had their own ups and downs, struggles and triumphs as they emerged from the North American wilderness to the world stage, and beyond.

Painting of pecan tree, leaf, and nut by Robert O'Brien. *Courtesy of Robert O'Brien.*

Pecans come packaged by Mother Nature in a lime-green husk that opens when the nut matures. Over time, the nuts fall to the ground or are shaken from the tree. *Photo used with permission of Georgia Pecan Growers Association.*

Pecans, a dietary staple of many Native American tribes, have been enjoyed by humans for millennia. The progress of this woodland tree nut's development into a commercial crop has had a close association with the history of the United States. *Photo courtesy of Kimberly Hatchett.*

Pecan kernels find their way into pecan pies, pralines, and an assortment of delicious recipes. *Photo courtesy of Patrick Conner.*

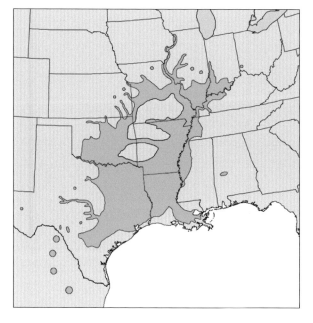

Left: Native range of the pecan tree in North America. *Map used with permission of USGS Geosciences and Environmental Change Science Center.*

Below: View overlooking the Mississippi River valley, near Vicksburg, Mississippi. Pecan trees grow along the high, well-drained ridges in the bottomland forests along the river. *Photo by Lenny Wells.*

A 160-year-old pecan tree growing on the grounds of George Washington's Mount Vernon Estate. The tree was removed in January 2014 because of risk of damage to the mansion. George Washington had a fondness for the pecan and was one of the first people to grow pecan trees in the United States outside their native range. *Photo by Lenny Wells.*

Lamar Jenkins budding pecan trees in a Georgia pecan nursery. Budding and grafting are methods of propagating or cloning individual plants. This practice was used by early pecan enthusiasts to bring about uniform crops of the same pecan from multiple trees. Application of this technology to the pecan led to the pecan's development into a major agricultural commodity. *Photo by Lenny Wells.*

Left: Spring foliage buds developing on the branch of a pecan tree. *Photo by Lenny Wells.*

Below: A pecan orchard during spring bud break, a splendorous sight throughout the southern United States. *Photo by Lenny Wells.*

Former National Champion pecan tree growing in Weatherford, Texas. The tree has a trunk diameter of 8 feet, a crown spread of over 159 feet, and a height of 118 feet.

Crimson clover growing in a pecan orchard in Crisp County, Georgia. Clover is grown in many commercial pecan orchards to provide supplemental nitrogen for the trees and foster a healthy soil. *Photo by Lenny Wells.*

Tuber lyonii, the pecan truffle, can sometimes be found growing in pecan orchards. The fruiting bodies of a mycorrhizal fungus that helps extend the pecan's root system, pecan truffles can bring in more than $100 per pound from specialty restaurants. *Photo courtesy of Timothy B. Brenneman.*

An example of the incredible diversity found within the pecan. There are over 1,000 recognized cultivars or varieties of pecan. *Photo by Lenny Wells.*

Left: Developing foliage, female flowers, and catkins, or male flowers, of pecan. Catkins bear the pollen, which is dispersed by wind to the female flowers. Each individual catkin stalk may hold as many as 220,000 pollen grains. *Photo courtesy of Patrick Conner.*

Below: Developmental stages of pistillate, or female, flowers of the pecan tree. *Photo courtesy of Patrick Conner.*

Squirrels are one of the primary predators and dispersal agents of the pecan.
Photo courtesy of Harry Bowden.

Left: Crows are the other primary predator and disperser of the pecan nut. *Photo courtesy of Matthew Hunt.*

Below: Many plants and animals call pecan orchards home, such as this green tree frog hidden within a cluster of developing nuts. *Photo courtesy of Rad Yager.*

Left: Developing pecan nuts showing symptoms of pecan scab, the primary disease of pecan wherever it is grown in humid climates. *Photo by Lenny Wells.*

Below: Modern air-blast sprayer used to spray pecan trees for pecan scab and other diseases, insect pests, and foliar nutrition. *Photo by Lenny Wells.*

The process of artificial pollination of pecan flowers is delicate. The female flowers are enclosed in sausage casings with a plug of cotton at the base of the bag. A hypodermic needle is inserted into the bag through the cotton plug to puff pollen onto the receptive female flowers. *Photo courtesy of Patrick Conner.*

Left: Pecan kernels packaged for use on NASA's Apollo space flights to the moon. *Used with permission of the Smithsonian National Air and Space Museum.*

Below: A pecan shaker used to dislodge nuts from the trees for harvest in the fall and to thin fruit from the tree in the summer. *Photo by Lenny Wells.*

Left: Orchard hedging machine for hedging pecan trees, a practice that is commonly used in arid pecan production regions, but which has recently become popular in the southeastern United States as well. *Photo courtesy of Tom Stevenson.*

Below: Mechanically hedge-pruned pecan trees, Stahmann Farms, New Mexico. *Photo by Lenny Wells.*

Young pecan trees planted in Sonora, Mexico. Mexico is estimated to have over 200,000 acres of pecan orchards, with only two-thirds of these old enough to be in production. *Photo by Lenny Wells.*

Trees of varying ages in an orchard contribute to a greater diversity of organisms associated with that orchard. A typical tree provides a predictable source of mating sites, plenty of options for egg-laying locations, refuges for hiding or resting, and a range of microclimates for a host of different species. *Photo by Lenny Wells.*

Pecan trees grow wild along the banks of the Colorado River near San Saba, Texas, known as the "The Pecan Capital of the World." Pecans have been a cash crop in the San Saba area since as early as 1857. *Photo by Lenny Wells.*

Pecan harvest at Pecanita, Brazil's largest pecan farm, near Cachoeira do Sul, Brazil. *Photo by Lenny Wells.*

Pecan orchard planted in China, Hunan Province. *Photo courtesy of Leonardo Lombardini.*

Pecans, known as *bigenguo* in the Mandarin language, are packaged for sale in China. *Photo by Lenny Wells.*

Visitors at a pecan booth during an International Food Exposition in Shang-hai, China, taste pecans for the first time. China's growing interest in pecans following 2007 changed the course of the pecan's future and breathed new life into the commercial pecan industry. *Photo by Lenny Wells.*

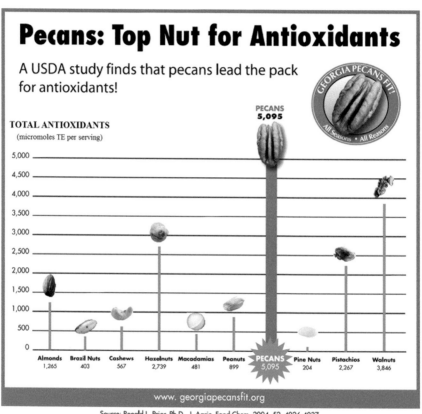

Pecans: Top Nut for Antioxidants

A USDA study finds that pecans lead the pack for antioxidants!

GEORGIA PECANS FIT!
All Seasons • All Reasons

**PECANS
5,095**

TOTAL ANTIOXIDANTS
(micromoles TE per serving)

	Almonds	Brazil Nuts	Cashews	Hazelnuts	Macadamias	Peanuts	PECANS	Pine Nuts	Pistachios	Walnuts
	1,265	403	567	2,739	481	899	5,095	204	2,267	3,846

5,000
4,500
4,000
3,500
3,000
2,500
2,000
1,500
1,000
500
0

www. georgiapecansfit.org

Source: Ronald L. Prior, Ph.D., J. Agric. Food Chem. 2004, 52, 4026-4037

Research continues to uncover a growing list of health benefits related to eating pecans.

Chart used with permission of Georgia Agricultural Commodity Commission for Pecans.

Pecan tree nursery production in the United States increased following the post-2007 demand for pecans on the world market. *Photo by Lenny Wells.*

Pecan trees being dug from a nursery near Valdosta, Georgia. Pecan tree planting has escalated rapidly since 2010. *Photo courtesy of Patrick Conner.*

· 5 ·

A Tree without Borders

As the twig is bent, the tree inclines.

—Virgil

THE HISTORIC TIES BETWEEN the pecan and the North American continent include Mexico, too. The pecan's native range extends to Oaxaca, some 700 miles south of the US border. Here pecans have been grown for ages.

Gilbert Onderdonk's 1911 report of pecan plantings in Bustamente, Mexico, estimated the trees to be 200 years old. If this age is substantiated, it would make them the earliest-known pecan trees under cultivation. However, Onderdonk originally judged orchards in Nuevo León to be 100 years old but later revised this estimate to only 60 years old.[1] Other reports suggest that during the Spanish conquest of Mexico more than 300 years ago, the orchards of Aguascalientes in central Mexico were already famous for their peaches, grapes, figs, and pecans.

Today, Mexico is the second-leading producer of pecans in the world, following the United States. The nut is now grown in 14 Mexican states, with most commercial orchards in Chihuahua, Coahuila, Durango, Sonora, and Nuevo León. As in Texas, early development of the Mexican pecan industry began with the management of native groves along the river bottoms, which were top-worked to improved varieties early in the twentieth century. The first commercially planted orchards of improved varieties, including Stuart, Success, Frotscher, Van Deman, Moneymaker, and Burkett, were planted in the early 1900s, primarily in Nuevo León, and were composed of trees grown from

Texas and New Mexico nurseries. While the plantings around Nuevo León mark the first attempt at modern commercial pecan production in Mexico, many of these early plantings failed because of the lack of specialized knowledge concerning pollination, soils, and adaptable varieties.[2]

Pecan production in Mexico as we know it today began with new commercial plantings in the 1970s of Western Schley, Wichita, and Mahan. Initially, alfalfa, watermelons, cantaloupes, beans, and corn were interplanted between the tree rows of some of the smaller farmers to offset the cost of orchard establishment. Later, alfalfa was abandoned in favor of the more shallow-rooted crops. Much of the success of modern pecan production in Mexico can be attributed to the efforts of Ruben Castro Medina, the longtime director of CONAFRUT, the cooperative arm of the National Fruit Committee. In the 1970s and 1980s, as a director of CONAFRUT in northern Mexico, Castro supervised over 40 field agents assisting pecan growers with every aspect of pecan culture. He developed and shared a wealth of knowledge on pecan propagation, varieties, management of cotton root rot, pruning, irrigation, and harvest management to help build Mexico's pecan industry.[3]

Castro maintained good working relationships with researchers and extension personnel in Texas, which proved beneficial for all parties. Castro demonstrated the benefits of deficit irrigation in Mexico, a management strategy that limits irrigation during noncritical periods of crop development, long before it was used in most areas of the world. Castro's abiding interest in and curiosity about pecans allowed Mexico to utilize the best production practices available to develop a burgeoning pecan industry almost overnight. Prior to the 1970s Mexico had no significant pecan industry, whereas today its production rivals that of the United States.

Like much of northern Mexico, Nuevo León has a semiarid climate that dictates that all serious commercial orchards be irrigated from rivers, lakes, or wells. During the 1970s, Nuevo León accounted for about 25 percent of Mexican pecan production. Although Nuevo León does not produce such a large percentage of the Mexican crop today, it remains a significant area of pecan production.

From Nuevo León, pecan cultivation spread westward during the 1970s into the state of Coahuila, which is today responsible for about 20 percent of Mexico's pecan production. Much of this production comes from the area known as the Comarca Lagunera, or the "Laguna District," a region surrounding the three sister cities of Torreón in Coahuila and Gómez Palacio and Lerdo in Durango.[4] This region is located in the central part of northern Mexico in the valleys formed by the Nazas and Aguanaval Rivers. The surrounding land is mountainous, at 3,700 feet in elevation.

In 1945 pecans were planted in the Comarca Lagunera along irrigation ditches used for other crops. According to Brison, these plantings included the varieties Ideal, Moore, Elliot, Hastings, Texas No. 60, Squirrel's Delight, Clark, Burkett, and Western Schley. These plantings were largely left to thrive or die on their own and did not lead to any significant pecan production. During the planting boom of the 1970s, however, pecan production in Coahuila began in earnest and was soon dominated by Western Schley, Wichita, and Mahan.[5] In 2003, there were a reported 35 pecan producers in the Comarca Lagunera region, managing just under 2,000 acres. Most of these were small producers with fewer than four acres of pecan trees, while only six producers grew more than 100 acres of pecans.

In this semiarid climate, water remains the key to producing pecans. About 60 percent of the orchards in the Comarca Lagunera are irrigated with well water. As in the arid production regions of the southwestern United States, soil salinity is a major production constraint in the Comarca Lagunera. Most of the region's orchards are irrigated by either flood or drip irrigation. The region has historically received about 20 inches of rainfall, most of which comes in August and September, greatly benefiting the quality of the region's pecans. Coahuila and Durango together harbor nearly 50,000 acres of pecan orchards, with 75 percent of those occurring in Coahuila.[6]

The development of pecans as a commercial crop in Sonora has made it the third-ranked pecan-producing state in Mexico, with over 21,000 acres of pecans. The first commercial pecan orchard in the desert region of Sonora was planted in 1952 by Luis Quiroz Freaner and Miguel Quiroz Freaner at the San

Rafael Ranch near the city of Hermosillo. Other significant early plantings in the region were made by Carlos Baranzini and Enrique Mazon, among others. As in much of the remainder of Mexico, the cultivation of pecan was expanded in Sonora during the 1970s, and a Growers Cooperative was developed around Hermosillo to aid in the planning of the state's pecan production.

With its northern border lying just below New Mexico and West Texas, the Mexican state of Chihuahua accounts for over 60 percent of the country's pecan production. The first commercial plantings in Chihuahua were set in 1946, near La Cruz, with the planting of 1,500 Western Schley and Bradley trees. Western Schley continues to be the most popular variety in Chihuahua because of its wide adaptation, precocity, production, and market acceptance.[7] Wichita and Mahan are also commonly planted here. Many new plantings in the late 1980s and early 1990s have now formed the core of Chihuahua's pecan production. Chihuahua is currently home to an estimated 131,000 acres of pecans. Over the next 10 years, this number is expected to grow since only around 75 percent of Chihuahua's orchards are old enough to produce a crop, as yet.[8] The native range of the pecan extends into Chihuahua and it can be found growing along the valleys of the Florido, Conchos, and San Pedro Rivers, which flow northward to the Rio Grande. Although commercial pecan production in Mexico comes primarily from improved varieties, native pecan trees have long been revered south of the US border, and it is of these trees that some of the most romantic stories of pecans are told.

Native pecan trees are said to have been cultivated in the state of Chihuahua for at least 400 years. The most lasting relics of these early trees are found in the town of Valle de Allende, established in 1563 near the Allende Valley. Here the land is rich with stories of the magnificent trees that grow in this fertile garden valley. Formed by the Rio Florido, a tributary of the Rio Concho about 385 miles south of El Paso, the valley was once famous for its Golden Delicious apples.[9] Although Valle de Allende is still home to a group of living monuments, some of its most magnificent specimens now live only in the memories and stories of the region's people. The valley's people have a tradition of naming their treasured trees: La Poza, San Miguel, El Colorado, and

Cracked. The most famous and grandest of these was "El Nogal de Músico" or "The Tree of Music," located at the entrance to the neighborhood of Paseo Jamaica. In 1941, the tree was described as having a limb spread of 150 feet and a height of 160 feet. Its diameter at the time was an incredible 10 and a half feet at four feet above the soil line, with a circumference of 41 feet. An old photograph shows 10 village children holding hands with outstretched arms encircling the tree. According to its owner at the time, Señor Gaspar Corral, the tree had produced an average yield of one ton of pecans per year and 200 successive crops. When questioned about the tree's age, Corral responded, "How old? No one knows."

By 1984, El Nogal de Músico was virtually dead, badly damaged by a series of lightning strikes, hollowed out in the center, with only a few living branches left as the tree struggled to survive. Today, the tree no longer exists, having finally succumbed to its last lightning strike. Despite its slow death, this massive tree remains a legend and is revered by the people of this small village. The tree acquired its poetic name many years ago when people would gather to dance under its widespread canopy. Some said that as the wind rustled through its leaves, the branches rubbed together, creating the sound of music like a violin's bow being played across its strings.

Another deceased giant was named "Sixto" or, "the Sixteen," which lived to the ripe old age of 350 years, according to locals. At one point, Sixto was struck by lightning, causing a large branch to fall. The fire ignited by the flash reportedly burned for eight days. When the fire was finally extinguished, the tree's holes were covered with stones. Over time the tree's bark grew over the stones, concealing them from view. When Sixto finally died, its massive limbs were sold for firewood. As the tree was cut up, the stones damaged multiple chain saws, adding a final chapter to the tree's colorful history.

Among the remaining legendary trees of the Allende Valley, "The General" stands full of life, hovering in a garden over dozens of surrounding trees. According to legend, General Pancho Villa kept some of the tree's nuts as he passed through the valley. Another "young" tree, estimated at only 150 years and called "Fat Boy," stands in a backyard with its roots surrounded by lilies.

"El Kilo" measures over 90 feet tall and obtained its name through the heavy nuts it produces. "Yesterday" is named in honor of the stories told under its branches by the elderly couples who have gathered to talk at the site under the tree for over 50 years. Perhaps the tallest of the remaining trees is "Solis," named for its owner, Refugio Solis Romero. Planted in a courtyard surrounded by other pecan trees, Solis stands 117 feet above the ground and is estimated at 200 years old.[10]

Today, pecan trees are planted in orchards throughout the Allende Valley with the primary goal of producing and selling pecans. But in this small village of Valle de Allende, pecan trees mean something more than money. They are revered as the stuff of legend and are accompanied by magical stories that can be heard in the rustling leaves of the remaining trees that continue to grow there.

Overall, pecan production in Mexico is booming. The country is second only to the United States in world production, harvesting a little over 250 million pounds of pecans in 2009. This production is likely to continue to grow, with an estimated area of over 200,000 acres planted to pecan and little more than two-thirds of this acreage currently producing nuts. Planting in Mexico continues at a steady pace at a reported 8,000 to 12,000 acres a year. It appears that Mexico's pecan production will be limited only by the availability of water for the crop. In 2011, producers with water wells on their property were the only farmers still planting pecan trees.[11]

Despite its proven economic value in North America, the pecan was not tested in the soils of other continents until sometime in the late 1800s, when North America's most valuable tree nut made its way to the Southern Hemisphere. The first Brazilian pecan planting is of particular interest because of its intriguing ties to largely forgotten events in US history.

Following the Civil War, the dejected, destitute, and defeated people of the Southern states were trying to make sense of what had just happened and what might lie ahead for them. The war had claimed over 600,000 American lives and left the land and economy of half the nation in a tattered shambles.

For many former Confederates, it was simply more than they could bear. So, they took up their families and left the United States entirely despite pleas by General Robert E. Lee that they remain at home until times got better. Some went to Mexico, some went to Cuba. As many as 9,000 of these newly disenfranchised people settled 5,000 miles south of the Mason-Dixon Line, in Brazil, with the hope of finding once again the life they had lost.[12]

Brazilian Emperor Dom Pedro II was interested in expanding cotton cultivation in Brazil and taking advantage of a worldwide depletion in the cotton market. He hoped to entice experienced cotton planters with knowledge of the latest agricultural practices, tools, and techniques. In return, he offered cheap land (22 cents an acre), tax breaks, land grants, and free passage to those willing to give this venture a shot.[13] The Confederados voyaged to Brazil in waves much as their ancestors had when they first came to America. Most would today be considered upper middle class—doctors, lawyers, teachers, engineers, businessmen, and of course, farmers. They shipped out of Galveston, New Orleans, Baltimore, New York, Newport News, and Mobile.

The journey itself was a gamble, and for most, rather unpleasant. On board ship, hammocks usually slept three. The diary of one colonist, Sarah Bellona Smith Ferguson, recorded an account of her trip to the new homeland aboard the ship *Derby* in 1866. The *Derby*'s Spanish captain was bribed by northern agents to wreck the ship in dangerous waters off the coast of Cuba. The plan was almost thwarted when the voyagers got wind of the plot and held the captain at gunpoint, ordering him to steer them through to safe waters. They were too late, however, and the ship slammed into the rocks. Fortunately, the passengers escaped and eventually made their way to their chosen destination.[14]

Gradually, several colonies were established: one in northern Brazil, 500 miles from the mouth of the Amazon River at the present-day city of Santarém, and another near the coast at Rio Doce. In southern Brazil, Juquia, New Texas, and Xiririca were established. Of all the colonies established by the Confederate expatriates, the largest, the most successful, and indeed, the most enduring, lies just outside the town of Santa Barbara, 80 miles northwest of São Paulo.[15] Here, the Confederados established a village dubbed Villa

Americana by the local natives. Eventually, in the mid-1930s, the community was officially named Americana.

The colony of Americana was established under the guidance of Colonel William H. Norris, a former soldier in the Mexican-American War and a senator from Alabama. Norris, like many of the Confederados, was a Freemason. As legend has it, during the Civil War, Union soldiers marching through Alabama stopped to raid Norris's home in Perry County, where Norris was said to have a small fortune in gold buried in the yard. It was reportedly a secret Masonic handshake delivered by Norris's wife to the Union officer in command of the troops that prevented the soldiers from taking Norris's gold. This gold was later able to buy 500 acres in Brazil from another Mason, Dom Pedro II, to establish the colony that would become Americana.[16]

Between 1865 and 1875, an estimated 2,000 to 4,000 US Southerners migrated to Brazil. While the Confederados were successful at establishing cotton as a major crop there, they introduced other crops to the new country as well, including watermelon, peaches, and pecans. While none of the pecan trees planted by the Confederados led to significant commercial production, they provided the colonists a taste and a reminder of the lives they had enjoyed at home before the war. Pecan trees planted by the Confederados remain today in Brazil as reminders of these expatriates.

Every July fourth, near the village of Santa Bárbara d'Oeste, about two hours north of São Paulo, girls dressed in colorful hoop skirts dance with young men dressed in Confederate gray beneath unfurled rebel flags at a site called the Campo, the remnants of the colony of Americana. Surrounded by sugarcane fields, the Campo consists of a cemetery, a chapel, and a monument. The Associação Descendência Americana, a society of Confederado descendants, gathers annually to feast on such Southern staples as fried chicken, coleslaw, and "pudim de banana."[17]

Many of the original Confederado immigrants returned to the United States. But those who stuck it out integrated into Brazilian life and managed to keep some of their own traditions alive in the process. Of the 161,000 residents of Santa Bárbara d'Oeste today, only about 30 families are of

American descent. The keeper of these traditions and the acknowledged expert on the history of the Confederados in recent times was a woman named Judith MacKnight Jones, whose great-grandfather Calvin MacKnight brought his family to Brazil. Jones's house, on land once owned by William Norris himself, is now surrounded by an urban landscape.[18] Amid the many changes to the landscape that have occurred since the Confederados made their way to this new land, a living monument remains in the form of a towering pecan tree shading the house. This seedling pecan tree, planted from nuts brought from Texas, serves as a living link to the place the Confederados left behind and the new life they hoped to create.

Brazil's commercial pecan production today has virtually nothing to do with the Confederados, however. The largest pecan orchard in Brazil, a 1,482-acre orchard near the town of Cachoeira do Sul in the southernmost state of Rio Grande do Sul, was the first large-scale commercial pecan operation in the country. The orchard is now owned and managed by Claiton Wallauer, whose family also grows rice, soybeans, and cattle. Mr. Wallauer's company, Pecanita, is the primary supplier of pecans in Brazil. Most pecans grown in Brazil are sold in the domestic market and are used by bakeries for candy and other purposes, with additional shelled pecans for retail sale and also for making pecan oil. The Pecanita orchard was planted in the late 1960s by previous owner Geraldo Linck under the advisement of former Auburn University horticulturist Dr. Harry Amling. Originally planted on a 30 foot × 30 foot spacing, the orchard consists of a mixture of more than 30 varieties, including Desirable, Cape Fear, Elliot, Barton, Shoshoni, Chickasaw, Shawnee, Western Schley, and Wichita. The orchard is planted on a relatively heavy clay soil and is mostly nonirrigated at this point.

The climate of southern Brazil is similar to that of southern Georgia, and the greatest challenge to growing pecans is pecan scab. For the first few years, Mr. Wallauer battled scab with a pair of old citrus sprayers, which were modified by the orchard's original owner to help the spray reach the tops of the larger trees. To accomplish this, the fans and booms were mounted on towers at the rear of the sprayer. Pecan weevil, hickory shuckworm, and yellow aphids also provide challenges. One of the most difficult pests with which Brazilian pecan growers must contend is the monk parakeet. While beautiful, these birds are the crows of Brazil, flying from tree to tree in flocks, cracking open pecans with powerful beaks and knocking them from the tree. In addition to birds, disease, and insects, crowding is one of the greatest challenges facing production in an orchard planted 30 feet × 30 feet. Until 2009, harvest had been done exclusively by hand at Pecanita. Today, mechanical shakers are used to remove the nuts from the tree and mechanical harvesters are used to harvest nuts on the most level ground in the orchards. Still, because there is some planting on terraces and unlevel ground, a portion of the first harvest of the year is done by hand as well. As many as 150 workers are employed for the second harvest, which is done entirely by hand. After harvest, the pecans are brought to the Pecanita cleaning plant, where they are cleaned and dried in large brick bins heated by a wood-burning steam engine.

Pecanita also has a pecan nursery producing Desirable, Barton, Shoshoni, Cape Fear, Shawnee, and Chickasaw varieties. The trees are grown in small pots under shade cloth in a medium containing a mix of sand and waste from the pecan cleaning plant. In order to build the volume of Brazil's pecan industry, Pecanita is encouraging the planting of pecans by small farmers looking for an alternative crop. While Brazil's pecan production is relatively minimal early in the twenty-first century, its pecan market is growing and with suitable varieties and management, Brazil's future with pecans is bright.

One of the most rapidly growing pecan regions in the world lies in Argentina. Pecans were introduced to Argentina in 1868 by Domingo Faustino Sarmiento, a legendary figure in the history of that country. Born in Carrascal, a poor suburb of San Juan, Argentina, Sarmiento rose from a rural schoolteacher to travel the world and become, among other things, an activist, writer, statesman, and eventually the seventh president of Argentina. Although his term has been marred by allegations of ill treatment of Argentina's indigenous people and his presidency was by and large considered a disappointment, Sarmiento remains one of Argentina's most well-known historical figures. Sarmiento obtained pecan nuts on one of his visits to the United States and disseminated them in Argentina. Several old specimens of these pecan trees can still be found in the provinces of Buenos Aires and Entre Ríos, suspected to date back to Sarmiento's period.

Around 1918, a group of Englishmen established plantations in the Paraná River Delta, on which pecans were planted. Some of these operations would later become part of the Tigre Packing Company, which, along with the University of La Plata, would supply seeds to Argentine nurseries in the early 1950s. Shortly thereafter, a National Institute of Agrotechnology agronomist named Martin Leber was instrumental in distributing pecans throughout the islands of the Paraná Delta. Thanks to his efforts, pecan trees can now be seen throughout the lower Paraná from the city of Tigre to the Uruguay River.

In recent years, pecan orchard establishment in Argentina has exploded to around 12,000 acres, although most of these are not yet in full production. As of 2012, there were about 400 pecan producers in Argentina. Over 90 percent of these were small growers with 25 acres or fewer. From 2009 to 2011, approximately 2,500 acres of pecans were planted each year. Much of the commercial orchard establishment to date has been in the northeastern and central regions of Argentina. Planting is expected to extend into the southern region including the provinces of Buenos Aires, south-central Mendoza, La Pampa, and Río Negro. Northwestern Argentine provinces are also making plans to expand pecan acreage in their region. In all, Argentina is expected to have more than 37,000 acres of pecans by 2020 if current planting rates continue.[19]

Pecan production has become popular in Argentina for several reasons. Aside from the growing demand for pecans and recent good market prices, pecan production creates a demand for labor and intensive use of the land. The longevity of pecan orchards is highly desirable to the Argentine people as well. Large pecan plantings were underway in 2012 throughout Argentina. In the Delta Entrerriano on the banks of the Paranacito River, nearly 2,000 acres of a planned 5,000 acres were planted to pecans by 2012. Plantings ranging in size from 150 to 800 acres are in various stages of planning and establishment from the delta of Buenos Aires to the province of Misiones. These acreages are made up of small investors who have never before been involved with agriculture, along with larger companies and small producers. The trees that supply these plantings are being produced in Argentine nurseries, which cooperate to provide varieties suitable for both the humid northeast production areas as well as the arid northwest. Trees for southern Argentina are producing varieties adapted to colder climates since the production region of southern Argentina is analogous to the northern US pecan region.

Current plantings in the humid northeastern region of Argentina include Stuart, Desirable, Sumner, Pawnee, Gloria Grande, Cape Fear, Forkert, Oconee, and Kiowa. Initially, Harris Super, Shoshoni, Success, Starking Hardy Giant, Mahan, and Kernoodle were used as well, although the popularity of these varieties has declined. Stuart, Mahan, Shoshoni, Western Schley, and Wichita are planted in the arid northwest, while in the colder regions of southern Argentina, Osage, Lucas, Colby, Starking Hardy Giant, Hodge, Major, Giles, and Peruque are used.[20]

With its abundant water resources, excellent soils, and suitable climate, Argentina could potentially become one of the world's major suppliers of pecans. In November 2013, Argentina was able to make its first shipment of pecans to China. The agricultural potential of Argentina, and for that matter, Brazil, is astounding. Their natural resources and climate have never been in question. Historically, Brazil has been limited by a lack of technological development, while Argentina has suffered from political and economic instability. However, since 2000, much on those fronts has changed and many South American countries, including Brazil, Argentina, Peru, and Uruguay, are expanding their agricultural horizons with pecans.

Although Colonel William Stuart shipped pecans from Mississippi to Australia in 1890, no one knows for certain when the first pecan trees were planted on the world's smallest continent. It is certainly possible, and highly likely, that some of the nuts sent by Colonel Stuart were planted, but the trail runs cold and there is no record of such an event. There are, however, pecan trees growing in Australian towns that were planted early in the twentieth century.

Many of these trees likely came from either the Langbecker Nursery at Bundaberg, Queensland, or the Petrie Nursery near Brisbane.[21]

According to Fred Brison's book *Pecan Culture*, one of the first commercial orchards in Australia was planted by Norman Greber near Gympie. Greber is renowned among nut producers in Australia for his pioneering grafting techniques with macadamia nuts. In 1955, Greber planted another small pecan orchard at Beerwah, Queensland. The first person to thoroughly commit himself to full-time pecan production in Australia, however, was A. T. Doyle, who planted 350 trees near Dagun, Queensland, including Stuart, Mahan, Western Schley, Pabst, and Williamson cultivars.[22] Because of the mild Australian winters, pecan trees often grow alongside citrus, papaya, avocado, banana, and pineapple, although today most pecan trees in Australia are isolated trees growing around homes and small orchards. It was not until the late 1960s that the largest commercial pecan orchard in Australia was planted by the same visionary family who brought pioneering pecan production practices to the southwestern deserts of the United States—the Stahmanns.

Around the world pecans thrive between latitudes 25° and 35° north and south. As April arrives and the buds of pecan trees in the United States and Mexico are popping open, pecan harvest is about to get underway in the Southern Hemisphere. Among the first to recognize the potential this reversal of seasons presented for growing pecans year-round was Stahmann Farms. Deane Stahmann Sr. and his sons had already established a highly successful pecan farming operation in the New Mexico desert. In the early 1960s, one of the Stahmann sons, Deane Jr., had worked for former senator and Republican presidential nominee Barry Goldwater, who would later lose the 1964 presidential election to Lyndon Johnson. Prior to that race, Deane Jr. had announced that he would leave the country if Johnson were elected president. After Goldwater's defeat, the younger Deane Stahmann was true to his word and convinced his father to purchase land in Australia. In 1966, Deane Jr. left the family's main operation in New Mexico to establish an 80-acre pecan orchard near Gatton, Queensland, at a farm called Las Piedras—a pilot project to test the waters of growing pecans Down Under. Stahmann quickly recognized the venture as a viable one and two years later, on a bend in the Gwydir River about 25 miles from Moree in the Australian state of New South Wales, he began planting pecan trees in earnest on a 1,850-acre farm, developing the largest commercial pecan operation in Australia, now known as Trawalla. By 1973, 72,000 pecan trees had been established.[23]

The Gwydir River Valley's black alluvial soil makes it one of the most agriculturally productive areas in the Southern Hemisphere. At Trawalla, Stahmann chose to plant the familiar varieties Western Schley and Wichita since the growing conditions closely resembled those of New Mexico. While the young trees were growing into production, Stahmann grew soybeans, sunflowers, beans, and cauliflower as cash crops between the pecan trees and on land not yet planted in trees. The trees were flood irrigated with water from the nearby river. In 1979, Stahmann's became one of the first pecan farming operations to sell pecans to China.

At Trawalla in the 1990s, Stahmann would dramatically advance Australian pecan production by experimenting with hedge pruning techniques, allowing Stahmann's to generate some of the highest pecan yields in the world—as much as 4,460 pounds per acre.[24] The dry Australian climate and excellent monitoring of imported graftwood have allowed the country to remain free of pecan scab. In addition, Stahmann's has pioneered pest management techniques using insect predators and fungi pathogenic to insects to develop an insecticide-free management system for pecans. In 2009, Las Piedras Farm at Gatton was certified organic and is now the largest certified organic farm in Australia.[25] A pecan nursery run by Stahmann's produces nearly 200,000 trees annually, supplying trees not only for its own orchards but also for those of other Australian producers.

In 2008, the Stahmann family sold their Australian pecan enterprise. The new owners retain the family name and continue the founder's strong commitment to the production and promotion of pecans. Continuing the tradition of sustainable management practices implemented by its founding family, Stahmann Farms has recently completed a $3 million project to convert 230 hectares of orchard from flood irrigation to drip systems, allowing them to cut their water use in half without reducing yield.

Australia now produces over seven million pounds of pecans on about 3,000 acres. Stahmann's supplies 80 percent of Australia's production on about 60 percent of the pecan acreage in the country. Thanks to the efforts of Stahmann's and the many smaller pecan producers of Australia, the pecan, virtually unknown in that country until the second half of the twentieth century, continues to grow in popularity.

Outside North America, no other place devotes more land to pecan production than South Africa. Here, at the end of the African continent, pecans have been cultivated for more than 100 years. First introduced to South Africa by a nurseryman named Wilkinson in the late 1800s,[26] pecans now occupy approximately 32,000 acres of land in this dynamic country, with only just over 30 percent of that currently made up of trees of bearing age. Average production in the country is expected to double about every third year going forward.

Initially, production was located in the Eastern Sector of South Africa. Much of this area is similar in climate to the southeastern United States, with hot, humid summers and mild winters. From the beginning, pecan production in this part of the world has been dominated by H. L. Hall & Sons, a large agricultural conglomerate. In 1890, Hugh Lanion Hall obtained a long-term lease from the Transvaal government for a large acreage of land near Nelspruit. Here, he worked to establish a productive diversified farming operation growing vegetables, fruit, nursery products, maize, and livestock. For many years, Hall & Sons' greatest strengths were citrus and vegetables. By the 1950s it was the largest citrus producer in South Africa. By the early 1960s, Hall & Sons had accumulated nearly 100,000 acres of land, allowing the company to expand its activities to include timber, tobacco, guavas, honey, game farming, irrigation engineering, property development, international marketing, and retailing. During the 1980s, avocados replaced citrus as the company's main export crop.[27]

Hall & Sons persevered through the political turmoil of the 1990s, which ended apartheid in South Africa. It remains a diversified farming operation, growing avocado, sugarcane, timber, pecan, and litchi, a stone fruit originating in China that resembles a cluster of hard strawberries with a sweet white flesh surrounded by a tough red shell. Hall & Sons planted the first commercial pecan orchard in South Africa in 1913 near Nelspruit in Eastern Transvaal. Moore, Barton, and Stuart made up the earliest pecan plantings. By 1980, Hall & Sons had planted approximately 1,100 acres with 55,000 pecan trees, about half of all the pecan trees growing in South Africa at the time.[28] Today Hall & Sons manages just over 600 acres of pecans.

In the mid-1960s, pecan varieties from the USDA breeding program, including Choctaw and Wichita, became popular in South Africa. However, because of the humid conditions, scab resistance proved to be a highly desirable characteristic for South African pecan production. In 1970, four local pecan varieties, Lane, Lerouk, Ukulinga, and Vlok, were selected by J. C. LeRoux in Natal for planting at the Fruit and Fruit Technology Horticultural Research Institute near Pretoria. All four of these local varieties originated as seedling trees in South Africa.[29]

The greatest boon to the South African pecan industry came from Nigel Wolstenholme, a professor of horticultural science at the University of Natal. Wolstenholme specialized in the ecophysiology of fruit and nut trees and became most recognized for his work on avocado. During the late 1960s and again in the late 1970s, Wolstenholme served on sabbatical at Texas A&M University, where he worked closely with USDA and Texas A&M scientists. Just as Ruben Castro had observed US pecan production practices and transferred that information for use by Mexican pecan growers, Wolstenholme observed and studied pecan varieties, spacing, irrigation, and other management practices, which he took back to South Africa.

Perhaps Wolstenholme's greatest contribution to South Africa's pecan production was in convincing the industry that pecans were being produced in the wrong areas. Although pecans have been grown commercially in South Africa for nearly as long as they have been in North America, the industry struggled to get off the ground. This was largely due to the planting of the wrong varieties in the wrong places, allowing pecan scab to take its toll. Having observed the effects of pecan scab in the United States, Wolstenholme knew that the humid eastern region of South Africa presented particular challenges to the crop's production. These challenges could be avoided by moving pecan production to the more arid regions of the country. As a result, the earliest South African pecan plantings in the humid regions of the former Transvaal Province near Nelspruit, White River, Tzaneen, and Louis Trichardt Districts have shifted largely to the more arid regions near Pretoria, the middle Orange River, and the Vaalharts. The most popular pecan variety produced in these arid lands is Wichita, which makes up nearly 60 percent of South Africa's pecan production today. The main factors limiting South African pecan production recently have been a shortage of nursery trees, the relatively low yield of pecans compared to other tree crops on valuable irrigated land, and irrigation water quality.

With the exception of the H. L. Hall & Sons operation near Nelspruit, the largest pecan plantings in South Africa are found today near Hartswater, at the center of the Vaalharts area, located in the Northern Cape and North West Provinces at the confluence of the Vaal and Harts Rivers. A semiarid region similar in climate to central Texas and Oklahoma, the Vaalharts has an abundant water supply thanks to the development of the Vaalharts Water User Association, which began as a government water plan managed by the South African Department of Water Affairs and Forestry. Organized in the early 1930s, the Vaalharts Water User Association became the largest of its type in the country, irrigating nearly 92,000 acres of farmland by 1976. Since then it has grown to incorporate other irrigation schemes and now provides water for more than 106,000 acres of crops, including cotton, deciduous fruits, peanuts, watermelon, barley, citrus, and vegetables, in addition to pecan. As in many arid regions, farmers relying on the water of the Vaalharts must manage the land for salinization to prevent the buildup of damaging chloride salts that can poison the soil. Thus, one of the country's main limitations to pecan production will remain its water management.[30]

While South Africa's pecan production may have languished for a few years, it is currently experiencing new growth, adding nearly 5,000 acres of pecans to the country each year. One of the largest producers in the Vaalharts region is De Meul Farming, with over 741 acres in production and as much as another 700 acres planned. Along with Wichita, the Pawnee, Choctaw, and Navajo varieties are commonly included in these latest plantings. In early 2014, big business invested in the South African pecan industry when Golden Peanut Company, a subsidiary of Archer Daniels Midland Company, which is based in Decatur, Illinois, purchased a 50 percent stake in South African Pecans, one of the fastest-growing pecan processors in South Africa, located in Hartswater.

The storied Holy Land of Israel is known for many distinctions. It encompasses sacred lands for both Christians and Muslims as well as for Jews. Pecans do not readily come to mind when we consider this country. Yet at one time pecans had a promising future in the Holy Land. While the events of this land extend back to before recorded time, the cultivation of pecans is a relatively new activity in this ancient place.

Pecan trees were first planted in Israel in the early 1930s, and throughout the following decade numerous varieties were introduced, including Delmas, Moneymaker, Burkett, Western Schley, Nelis, San Saba, Halbert, Garner, Big Z, Desirable, Govett, Schley, and Mahan. Limited commercial planting of pecan occurred until the late 1950s, when Delmas and Moneymaker became the predominant varieties. One of the largest commercial plantings at the time was around 250 acres. Initially planted to rootstock, the operation failed because of poor management, which deterred commercial interest in pecan for several years. However, high prices continued to fuel small-scale interest in pecans. Even as commercial interest languished, a total of about 1,500 acres of small plantings of three to five acres each were developed throughout the area. During the 1960s, as interest in pecan grew, Israeli growers ceased planting Moneymaker and replaced it with Burkett, Mahan, and Western Schley. Yet Delmas remained the most popular variety.[31]

More pecan varieties were introduced into Israel in the late 1960s, including many of the newly emerging varieties released by the USDA. Throughout the 1960s and into the 1970s, pecan planting continued at a rapid rate. By the spring of 1975, the total pecan acreage of Israel had reached a peak of almost

15,000 acres.[32] Israel's pecan production occurred mainly along the coastal area bordering the Mediterranean Sea and along the hot northern inland valley of the Jordan River, alongside citrus and bananas.

Early production problems resulting from varieties of low quality and the management of tree crowding have hampered Israeli pecan production. Most early commercial plantings were spaced 33 feet × 33 feet, which began to crowd and limit production within seven years or so. Additionally, the poor nut quality and severe alternate bearing tendencies of the Delmas variety began to surface once the trees reached maturity. Western Schley and Mahan suffered similar problems.

Israel has varied soil, climate, and topographical conditions that allow it to grow a wide range of crops. Compared to many of Israel's other orchard crops, pecans require little labor and generate a high return. One major advantage to pecan production in Israel is, as we see in most arid regions, an absence of disease, and very few insect pests. However, Israel's future pecan production faces many challenges. For one, Israel is a small country of only about 8,000 square miles, roughly the same size as New Jersey. Like the population of the rest of the world, Israel's population is growing. While its demand for agricultural products will increase, so will its urban use of land and water. Land availability for agriculture is at a premium in Israel and will likely decline. Perhaps no limitation is as great in this region as that of water. Pecans are a thirsty crop and Israel is a thirsty nation. The amount of freshwater allocated for agricultural use in Israel was reduced by half in 2000. It is estimated that by 2020 the water availability for agriculture may be even less.[33] Thus, the infrastructure required to support pecan production in Israel is seriously limited.

Since the development of drip irrigation by Simcha Blass, Israel has been a leader by necessity in pioneering new technologies to enhance water-use efficiency. One answer to Israel's water limitations in recent years has been desalination. The Israeli government's desalination program, SWRO (Sea Water Reverse Osmosis), was initiated in 1999. The program is designed to alleviate demands on Israel's scarce water resources, particularly in the face of drought conditions that have plagued it most years since the mid-1990s, by removing the salt from water supplied by the Mediterranean Sea. Since the initiation of the desalination program, the targeted annual quantity of desalinated water to be produced has gone up and down with short-term changes in rainfall and national consumption rates. The initial target capacity of 50 million cubic meters changed four times between 1999 and 2008 and currently rests at 750 million cubic meters. The program began contributing potable water to the national water grid in 2005. Today, desalinated water provides about half of Israel's drinking and agricultural water—more than in any other country.[34]

There is, however, some concern about the use of desalinated water on agricultural crops. When the salts are removed from seawater, required nutrients like calcium, magnesium, and sulfur are removed as well. Boron, on the other hand, is not removed by the desalination process, which can lead to problems for crops since the boron concentration of seawater is often high enough to be toxic to many plants. Thus, farmers using desalinated water need control systems to compensate for the poor water quality if minerals required for agriculture are not added or removed at the desalination plant.[35]

Currently Israel has only about 500 acres of pecans in production. But, as in the rest of the world, there is a renewed interest in pecan production here. While it is unlikely that Israeli pecan production will compete with US or Mexican pecan production, this small country keeps pecan culture alive in the Middle East.

While pecans are gaining in popularity, they are still largely unknown throughout much of the world. Yet even in the 1960s NASA was aware of the value of the pecan. It was searching for nutritious snacks for astronauts on the Apollo missions of that era. Space food had to be stored without refrigeration and had to have the ability to be eaten under weightless conditions, while being nutritious, lightweight, and capable of being compressed when possible.[36] These requirements led Houston's NASA engineers to pecans. They began preparing Desirable pecans from Texas A&M orchards for the various Apollo flight crews. Pecans were used on two flights to the moon, Apollo 16 and Apollo 17, in April and November of 1972. The pecan is ideal for this purpose since it can be dried to 3 or 4 percent moisture, will freeze down to -170° C, and will thaw without ill effects. It is completely digestible even in the raw state. On the Apollo flights, the pecans were protected with a four-ply laminated film coating. This protected the food from loss of flavor and moisture, as well as from oxygen invasion, spoiling, and excess crumbling. One of these pecan packs from Apollo 17 is now housed at the National Air and Space Museum in Washington, DC.[37] This value of pecans as a nutritious, portable food in the opinion of NASA would later become the nut's main selling point in areas of the world that were otherwise unfamiliar with the pecan.

· 6 ·

Healing the Land
with Orchards

I saw, and behold, a tree in the midst of the earth, and its height was great.
The tree grew and became strong, and its top reached to heaven, and it was visible to
the end of the whole earth. Its leaves were beautiful and its fruit abundant, and in
it was food for all. The beasts of the field found shade under it, and the birds of the
heavens lived in its branches, and all flesh was fed from it.

—Daniel 4:10–12

SINCE THE PUBLICATION OF Rachel Carson's *Silent Spring* in September 1962, concern over the sustainability of agricultural production systems has heightened. Agroecology, the application of ecological principles to the production of food and fiber, has grown in prominence nationwide. The goal of agroecology is to make crop production compatible with biodiversity in a sustainable manner.[1] Orchards innately, though not necessarily by design, offer a unique system by which this goal may be attained. In fact, some now believe that portions of the Amazonian rain forest, traditionally revered as pristine forest and the poster child of sustainability, are actually ancient orchards of peach palms, Brazil nuts, and a wide assortment of fruit trees planted and cultivated by the early Amazonian Indians.[2] Only when these forests or "ancient orchards" are cleared for annual crops are the fragile tropical soils denuded by exposure to tropical rains. The salvation of these lands is thus found in trees.

The potential for orchard systems, including pecan, to achieve the goal of agroecology is related to three distinct features: their permanency, their incorporation of other species, and the fact that a system selecting for trees provides multiple niches for other species to exist at various levels within a vertical plane. Together, these factors generate a remarkably healthy and sustainable system, even within the context of the "disturbed" environment of conventional agriculture.

Author and president of the Land Institute Wes Jackson describes soil as ecological capital. Years ago, he noticed that regardless of human efforts, soil on sloping ground eroded wherever annual monocultures were planted, but it would stay put in perennial pastures, native prairie grasslands, and forests. The prairie ecosystem, he observed, counted on species diversity and genetic diversity within a species to weather epidemics of insects and pathogens. In addition, the prairie maintained its own fertility and actually accumulated soil over the years.

In the southern half of the United States, pecan orchards can and often do accomplish these same ecological goals. One difference between the Southeast and the Midwest is that the former region was originally covered in forest, in contrast to the prairie that covered the latter. Thus, a system that mimics the forest system, as do orchards, should be a harmonious form of land use. In pecan orchards, this harmony is now both practical and profitable. Pecans are long-lived perennial trees, and when given an opportunity, pecan orchards will persist for more than 100 years. This persistency can have positive effects, literally from top to bottom. Such a permanent host plant and its associated cultural practices can enhance the sustainability and resilience of the entire system,[3] attributes that are true for both commercial, conventionally managed orchards and for organic orchards. Organic farming relies exclusively on ecological processes adapted to local conditions rather than on synthetic inputs like commercial fertilizer and chemicals. Conventional farming is often labeled as relying solely on commercial inputs for production and has received the stigma of doing so at the cost of a degraded environment. While there are certainly instances where such a stigma is warranted, there is increasing evidence that we can achieve a sustainable agriculture with wise use of practices pulled from both philosophies. Commercial pecan production can present a model system for sustainability.

The foundation of sustainability for both conventional and organic agriculture lies in the soil. The vegetative cover of natural ecosystems like forests or grasslands protects the soil, preventing erosion, replenishing groundwater, and enhancing the slow infiltration of water into the soil. In the southeastern United States, pecan orchard soils are normally left undisturbed for many years. While weed-free areas may be maintained along the tree rows of commercial orchards to eliminate competition for water and nutrients around irrigation lines, the area between tree rows remains covered in vegetation, primarily grass and legumes, to prevent compaction, hold soil in place, and provide a more dust-free and clean harvesting surface and environment. In all, roughly 70 percent of the pecan orchard floor remains vegetated. Even where weed-free strips are maintained, the trees themselves help protect and preserve the soil from the eroding effects of heavy rainfall. Thus, an orchard is essentially a meadow interplanted with rows of trees. Throughout the history of human land use, the healing of abused land has centered on trees, grass, and in many cases, legumes—all crops that have promoted permanency, enrichment, and protection of soil. All of these land-healing plants are commonly found in pecan orchards east of I-35.

Composed of a near-magical complex of bits and pieces of dead and dying organisms and nonliving rocks washed and pulverized over the eons by weathering, the soils on which agriculture depends are literally teeming with life. While there is much that we do not know about the secret lives of the plethora of minute bacteria, fungi, and other organisms that inhabit our soils, we do know that without them the soil on which we depend would be sterile, unable to support life, including our own. These organisms break down the land's plant and animal residue, converting it to soil. As they munch, mash, and digest the earth's natural litter, they mix it with the surface soil, enriching it anew. It is this intricate web of life and death that allows our crops, including pecan, to flourish.

While commercial pecan orchards are without question disturbed ecosystems managed with the primary goal of producing a crop and income for the producer, they are in many ways excellent mimics of nature's pattern. Most modern agricultural systems are productive only as a result of their dependence on external inputs like chemical fertilizers. The relative sustainability of nontilled orchard systems is evidenced by studies revealing that orchards lacking fertilizer application for up to six years suffer no drop in crop yield or tree nitrogen content.[4] For example, I once observed a mature pecan orchard in Texas on the banks of the Colorado River that had never been fertilized, yet it still bore a heavy crop and healthy, deep rich green foliage. As they enter dormancy each year, trees conserve many nutrients that they later use in subsequent years. The pecan tree's buffer against low fertility is fueled partly by reserves stored within the tree. But the resilience of orchard soils is fed largely by the recycling of nutrients over time, just as occurs in forest soils.

Trees in general are a vital piece of one of nature's most remarkable feedback loops. They grow and produce foliage to capture sunlight and convert it to energy used for more growth and reproduction. At the same time, the roots of the tree grow outward in a seemingly endless complex of branchings that spread to twice the width of the aboveground canopy in a constant search for water and nutrients to maintain the internal processes. The massive root system of a pecan tree alone makes up nearly a quarter of the tree's total dry weight.[5] In the process of intercepting sunlight, the leaves shade the ground beneath the tree's canopy, cooling the soil and reducing the loss of moisture so that the environment is more suitable for the proper functioning of the roots. As the heat of summer wanes and the days grow shorter, the tree's internal chemical changes lead to the leaves' separation from the branches. They fall to the ground below in bulk. The remaining nutrients in the leaves are released slowly over the course of several months to years, enriching the soil and improving its condition around the tree as they are decomposed by an array of soil organisms. The remaining energy in the tree is stored for use in subsequent years. We need look no further than the tree for a lesson in sustainability.

The key to this sustainability is soil organic matter. The many merits of organic matter were discussed in chapter 3. The soil, composed of various stages of living and dead or decaying plant and animal material, is the lifeblood of plants, which are fueled by the nutrients that are released in the process of decomposition. Decomposition enhances nutrient cycling, soil moisture retention, and soil formation. Depending on the environmental and climatic conditions in which they are found, soils can vary greatly in the amount of organic matter they can hold. In the southeastern United States, soil organic matter levels are often quite low compared with those in other regions of the United States. For instance, soils in the northern Great Plains of the United States have some of the highest organic matter levels of all mineral soils, commonly ranging from 4 to 7 percent. Most soils in the state of Georgia range from 0.5 to 2 percent organic matter. Yet pecan orchard soils in Georgia average 3.6 percent, a level comparable to that of the natural hardwood forests of the region.[6]

These organic matter levels are achieved, in part, through the deposition of leaves, bark, twigs, branches, shucks, and other debris that come raining down to the orchard floor throughout the course of the year. A single 15-year-old pecan tree is estimated to carry about 66 to 175 pounds of leaves alone.[7] This may not sound like all that many leaves until you consider the feathery weight of a single leaf. At a spacing of 12 trees per acre, 800 to 2,000 pounds of leaves would reach the orchard floor, where they would decompose and release their nutrients slowly over time to once again enter the orchard's nutrient cycle.

Because of their relationship with bacteria in the soil, legumes are able to harness nitrogen from the atmosphere for their own use. As the plants decompose, they leave the nitrogen behind in the soil for other plants to absorb. Thousands of years ago, the earliest farmers of Southeast Asia practiced crop rotation with legumes like soybeans, peas, and chickpeas. Historical references to the benefits of legumes can be found in the records of the Roman Empire, going back to Cato the Elder (234 to 149 BC), who suggested improving vineyard soils with legumes.[8] Although they had no conception of why or how it happened, these farmers and thinkers recognized the restorative effects that legumes had on the soil. In the 1880s, two scientists, Herman Hellriegel and Mikhail Voronin, independently discovered the process of nitrogen fixation by microorganisms associated with the roots of leguminous plants, uncovering the hidden pathway of the earth's nitrogen cycle.

The planting of legumes such as clover enriches the soil with nitrogen. As Hellriegel and Voronin discovered, the plants do this through the action of certain species of soil-dwelling bacteria in the genus *Rhizobium*. Only a specific species of *Rhizobium* can provide optimum nitrogen production for each group of legumes. Although *Rhizobium* bacteria occur naturally in many soils, when planting legumes, a farmer applies or "inoculates" the seeds with the bacteria, which come in small bags full of black powdery material that looks like soot. The bacteria colonize the roots of the legume plant as the appropriate bacteria come into contact with the root hairs. The root hairs encircle the bacteria, creating a nodule, a term used to describe a warty-looking lump on the root surface. The nodules range in size from that of a BB to a kernel of corn. As nitrogen gas passes from the air between soil particles and enters the nodule, the bacteria produce an enzyme that helps convert the nitrogen gas to ammonia, an element used by the plant. In return, the bacteria are supplied with the plant's sugars to fuel the process. The plant can expend considerable energy, as much as 20 percent, in fueling the bacteria's work. Therefore, if nitrogen is readily available in the soil, a plant will lazily utilize the available nitrogen rather than expend energy to feed the bacteria. If the bacteria are actively fixing nitrogen from the atmosphere, the interior of the nodule will appear pink upon slicing it open.

Aside from the wonders it can do for improving the soil, crimson clover is achingly beautiful when flourishing in full bloom amid the lime-green backdrop of young pecan leaves each spring. It is at this time that nitrogen fixation is at its peak. While they are alive, clover plants release little to no nitrogen into the soil. But after the seeds are formed and nitrogen fixation ends, the nodules slough from the roots, and the bacteria lie in the soil awaiting the next opportunity for colonization. In many pecan orchards, crimson clover is allowed to grow until May or June before mowing so that it can reseed itself. In years of good spring soil moisture, the clover may reach knee high before mowing. After the plants die and dry down, the nitrogen in the clover's roots, stalks, leaves, and seeds becomes available for utilization by other plants like pecan trees as the clover is decomposed by microbes, worms, and insects. The clover plants themselves add tremendous amounts of organic matter to the soil. Crimson clover alone can provide over 3,000 pounds of dead or decaying plant material to an acre of soil.[9] This material is combined with that of the tree's discarded parts, grasses, and weeds in the orchard to contribute greatly to the health of the orchard soil.

As the cost of fertilizer rose with increasing fuel costs in 2008, the use of legumes in orchards became common practice in some regions for the first time since World War II.[10] Many orchards in the southeastern United States are now filled with a colorful carpet of crimson clover and hairy vetch through the winter and early spring. Many also harbor Italian ryegrass and wild radish, escaped "weeds" that seem to thrive in this environment. Underlying this cover is a mat of grass, usually bahia or Bermuda, which grows thick and lush in the warm months after the winter cover has seeded out and been mown down.

While leguminous plants like clover and vetch are planted and encouraged to enhance organic matter and supply nitrogen, they perform other functions beneficial to the soil—namely, improving the soil environment for the myriad tiny creatures that further enhance soil health. A plethora of fungi, bacteria, and invertebrates thrive with elevated soil organic matter and help keep the soil system running smoothly.[11] The warm-season grass cover also helps protect the soil from erosion. Even the weedy wild radish helps break up compaction as the large roots grow through the soil, creating cracks and crevices that aid in filtering water down into the subsoil.

Soil is a dynamic force. It is constantly changing, improving, degrading, and at times stabilizing, depending on its history, use, abuse, and the environment around it. The extent to which it can be improved depends on the type of soil. The ancient, degraded soils of the southeastern United States are low in most of the nutrients required to support crop production. Many orchards of the region are planted on sandy or sandy loam soils worn out in the early twentieth century by successive crops of cotton. When orchards are being established on these soils, and at least periodically thereafter, one or more nutrients must be added to the system to prime the pump. But, thanks to nutrient cycling and the surprisingly small amount of nutrients removed from an orchard annually by the pecan crop itself, once nutrient levels reach an optimum, pecan orchards are buffered from the need for high nutrient inputs on a regular basis, even on the degraded upland soils of the southeastern United States.[12]

Under some conditions, overfertilization can be more common in orchard crops than in other crops because of the old practice of fertilizing during plant dormancy. The nutrient most commonly limiting crop production is nitrogen. Its importance is illustrated by the fact that over three-quarters of the nitrogen found in plant leaves is associated with the process of photosynthesis, without which plants won't survive. For orchard crops, the amount of nitrogen removed by the crop is far less than with more traditional crops. As a result, many pecan scientists now suggest reducing the traditional rates of nitrogen that have been applied for years to pecan. Annual crops are generally fertilized in split applications or when the crop itself can best utilize the fertilizer. Proper timing of fertilizer application is key to reducing the loss of nutrients, but only recently have scientists come to better understand how and when pecan trees use nitrogen.

After World War II, cheap synthetic fertilizers were commonly used on all crops, usually in the spring when the season's new growth began. However, farmers and fertilizer dealers could not apply fertilizers to all crops at the same time. Pecan tended to take a backseat to more "important" crops like cotton and corn. As a result, pecans began to be fertilized prior to budbreak in late winter or early spring because that was when most people had the time to do it, and for the most part the trees seemed to respond. Today, we know that this may not be the best time to fertilize pecan trees.

The energy required to support the initial flush of pecan leaves in the spring is supplied by the tree's energy reserves. Only after this supply is somewhat depleted does the tree's demand rise and its uptake of nitrogen from the soil increase.[13] As a result, most pecan producers in the southeastern United States now apply their spring fertilizer after the first flush of growth. If the application of fertilizer is timed to meet the tree's needs, there is less nitrogen lost to the environment through leaching.[14] Unless removed by plants, excess nitrogen in the soil eventually makes its way into surface waters and aquifers. While this loss may be negligible in arid regions, in humid regions such as the southeastern United States, it can be heavy. The pecan tree's roots, which are often deep and occupy an enormous volume of soil, act as a safety net to prevent such loss, removing nitrogen that may leach down deep into the soil, reducing groundwater contamination and increasing nutrient use efficiency.

Phosphorus, another major nutrient required by plants, is a common source of environmental pollution. Phosphorus tends to accumulate in soil and can find its way into surface waters through runoff and erosion. The pecan tree root system and constantly vegetated orchard floor hold the soil in place and reduce this runoff and soil loss.[15]

During the 1970s and 1980s, the use of broad-spectrum insecticides in orchards of commercially improved pecan cultivars was enormous. Pecans have an eight-month growing season, long by crop standards, and they are vulnerable to damage from various insect pests throughout that period. Organophosphate and pyrethroid insecticides were routinely used in the 1970s and 1980s and were often sprayed 8 to 10 times in a given growing season on a regular schedule. This led to a period of reliance on chemical control of pests, which dramatically increased yields but was a terribly destructive practice that wiped out beneficial insect populations as well, allowing a subsequent proliferation of many pecan pests, most notably aphids and mites.[16] A sort of catch-22 situation developed in which producers needed to control pests but used methods that provided only temporary control and led to more problems down the road. Today, pesticide and fertilizer use in productive, conventional commercial pecan orchards remains significant, although the levels are considerably reduced compared to those of a few decades ago. Management styles vary depending on the region and regime in which pecans are being grown in the United States. In the native stands and groves, pecan growers may fertilize their trees but rely on natural controls and very limited use of pesticides to manage pests. In the Desert Southwest, there is no need to apply fungicides in the absence of the free moisture conditions that drive pecan scab, but insect management remains a challenge and relies on chemical control. The same is true in the humid pecan-producing regions, where both fungicide and insecticide applications are required for successful production of certain varieties. As a result, agricultural chemical usage in US pecan orchards varies considerably between states and from grower to grower.

It has been estimated that pecan producers used approximately 333,000 kilograms of insecticide in Georgia and Texas during 2001. At the time, 69 percent of the US pecan acreage was treated with supplemental nitrogen fertilizer, 67 percent was treated with insecticides, 47 percent was treated with fungicides, and 44 percent was treated with herbicides.[17] While pesticides are still used today in commercial orchards, there has been an industry-wide shift to less reliance on chemicals and the use of less-harmful insecticides that target specific pests. This change has led to far fewer insecticide sprays and a greater reliance on beneficial insects and pathogens to help suppress pest populations below damaging levels. This form of integrated management requires regular monitoring of pest populations rather than indiscriminate spraying to control pests. In fact, many pecan producers employ seasonal scouts to keep watch for insect pests throughout the growing season. Some, like Richard Grebel, drive through the orchards in elevated stands mounted inside the bed of a pickup. Grebel and his small army of scouts, armed with small hand lenses and handheld clickers, keep a running tally of each insect pest they encounter. They scout 5,000 acres of orchards across Georgia in order to monitor population levels and help their clients make decisions about what and when to spray.

The warm, humid climate of the southeastern United States provides a hospitable climate for pecan trees. Unfortunately, it also provides a perfect environment for fungal diseases, most notably pecan scab. As a result, commercial pecan production from the Carolinas to central Texas would be much less productive in the absence of fungicide sprays. Producers in these areas who grow scab-susceptible varieties sometimes spend more money on fungicide applications than they do on any other aspect of their production. Unlike insecticides, fungicides are preventive. Therefore, the material must be applied to the tissue that needs to be protected prior to infection. This means that during rainy periods growers must apply fungicides on a calendar schedule, usually every two weeks. During dry periods, these intervals are often extended.

Many growers have found that cultural practices can help manage their disease problems and minimize the number of fungicide applications. One such practice is to remove trees from overcrowded orchards. As pecan trees age, they of course grow larger and take up more room, often shading their neighbors, blocking air flow, and trapping moisture beneath the dense canopy, which keeps the nuts and leaves wet for longer periods, allowing scab and other diseases to thrive. In these situations, the trees become considerably stressed and less productive. So, when trees in an orchard reach a certain size, they are removed to alleviate a large portion of the problem. Usually when orchards are thinned appropriately, everything about the orchard is enhanced. Not only is disease pressure reduced, insect pests are less of a problem, their management becomes easier, and the trees are more productive. Still, some pecan varieties are simply more genetically predisposed to disease than others, requiring fungicide applications, particularly during wet years. As a result of the growing expense of fungicides and fuel to spray them, many growers are choosing to simply plant varieties that are more disease resistant. Depending on the level of scab susceptibility the chosen cultivar has, a farmer can eliminate, or at least cut back significantly on, the number of fungicide applications required.

One of the most common gauges of the health of any ecosystem is its biodiversity.[18] Modern agricultural ecosystems are often considered biological deserts because of intensive human selection for a single species, which provides the crop being grown with a significant artificial competitive advantage. This is referred to as monoculture. However, among agricultural systems, perennial agroecosystems such as orchards can be and often are, whether intentionally or not, an exception to the modern rule. On the surface, a pecan orchard may appear to be a monoculture, but many other species can be located there.

Diversity or variety has been called the security of agriculture, and indeed of life itself. Not only can pecan orchards be diverse in the number of plant species present, they can also harbor a large amount of genetic diversity within what is superficially a pecan "monoculture." Unlike large fields of annual crops such as corn, cotton, soybean, peanut, or wheat, which consist of a single variety in true monoculture, orchards often consist of multiple varieties of pecan within a single orchard. Trees within native groves are diverse by nature, not only genetically but in time and space as well, because trees of varying sizes and ages are spaced randomly throughout the grove. Improved-variety orchards usually consist of at least two to four varieties in a uniform spacing. They are planned primarily for the purposes of pollination, but many growers choose diversity as a hedge against the potential failure of any one individual variety in a given year as a result of a late spring freeze, disease, alternate bearing, or other factors. In the early days of US pecan production, orchardists didn't plant a single variety, but multiple varieties. This practice led to the designation of "the big four" varieties planted in the South at the time—Stuart, Schley, Alley, and Pabst. One advantage to diversity is that it inadvertently minimizes pest pressure within an orchard because it requires a greater adaptation on the part of the pest in order to proliferate,[19] which gives nature time to make adjustments.

An excellent example of such foiling of pest infestation is found in the relationship of pecans and pecan scab. The virulence of the pecan scab fungus is largely specific to each variety. The more often the scab fungus is exposed to the right environmental conditions on the same variety, the more damaging the scab epidemics can become. The genetics of pecan scab and its ability to adapt to new varieties present particular challenges to pecan breeders looking to develop scab-resistant varieties. History has shown that new pecan varieties often enjoy a "grace period" following their introduction before scab can overcome their innate defenses. Some pecan breeding programs, like that of Patrick Conner at the University of Georgia, are seeking to take advantage of the incredible diversity of the pecan species itself, incorporating different resistance genes through classical breeding into new varieties that can be transitioned into orchards to maintain diversity. Pecan producers do something similar, on a much less technical level, when they simply plant a mixture of pecan varieties in their orchards. Theoretically, they can reduce disease, or at the very least slow the development of pecan scab epidemics.

There are those who will ask how an agroecosystem that receives multiple pesticide applications can be sustainable. Cultural practices have a dualistic relationship with biodiversity. Some, like pesticide spraying, can reduce or alter biodiversity. Others, like the planting of legumes and multiple crop varieties, contribute to the agroecosystem's richness. We mustn't forget that an orchard's productivity depends on the healthy functioning of the ecosystem itself. As long as our methods of crop production do not compromise the vital functions of the system but instead enrich them, orchards can thrive for everyone's benefit.

Understandably, the less pesticide used in an orchard, the greater the diversity of life that will be found there. The genuine integration of natural, cultural, biological, and selected chemical techniques has led to a 35 percent reduction in the amount of insecticides used in pecan orchards over the last 25 years.[20] Evidence that current pecan management practices can be sustainable is found in the diversity of life in the orchards. Despite pesticide use, the biodiversity in conventionally managed pecan orchards is astounding. Even in the midst of aerial fungicide applications, soil fungi, including the root-extending mycorrhizae fungi, thrive and form the base of food webs in nontilled agricultural systems like pecan orchards. Beneficial, harmful, and innocuous insects are found throughout the orchard from the soil through the understory and into the tree canopy, where appropriate insecticides are used responsibly.

In my position as pecan specialist at the University of Georgia Horticulture Department, I see hundreds of pecan orchards each year. In a season-long survey during 2012, I attempted to document each living species I encountered in conventionally managed Georgia pecan orchards. These included many species of plants, fungi, mammals, birds, insects, reptiles, mollusks, and amphibians. Of the 202 species of organisms I cataloged, only 40 percent were considered pests (insects, diseases, weeds, nut-feeding birds and mammals) when occurring in high populations. However, pests contribute to orchard biodiversity as well. My informal survey is in no way comprehensive. There were many species I could not identify, but evidence was good that pecan groves in the Southeast provide habitats containing a high degree of biodiversity.

There are, in fact, 180 species of pecan-feeding insects and mites.[21] Each of these in turn is associated with a wide array of natural enemies like lady beetles, lacewings, spiders, parasitic wasps, fungi, and nematodes. The use of pesticides can have a negative effect on many arthropods, including beneficial insects and spiders. However, total arthropod richness of the tree canopy is not affected or is very little affected by some pesticides since many are selective for specific pests or pest families. A Texas pecan orchard survey documented 181 species of spiders alone from 26 families, making the pecan agroecosystem one of the richest in spider fauna.[22]

Aphids are some of the most abundant insect pests in pecan orchards. One of the most effective groups of aphid predators is lady beetles. At least seven species of lady beetles are commonly found in pecan orchards. Most prefer feeding on aphids close to the ground. But one in particular, the multicolored Asian lady beetle, *Harmonia axyridis*, is an arboreal species, preferring to feed in trees. While the planting of cool-season legumes and other plants in the orchard often increases lady beetle populations, they may not necessarily be feeding on pecan aphids. Most of the lady beetles fail to move up into the trees to feed on the aphids if they find suitable prey closer to the ground. However, when the Asian lady beetle population is increased, so also is biological control.[23]

Introduced initially in 1916 by the USDA, and several times thereafter in various locations, the multicolored Asian lady beetle first became established in Louisiana near the port of New Orleans. Thus, it probably gained a foothold in the United States via a series of planned releases and accidental entries. While North American lady beetles have specific color patterns, the Asian species is quite diverse in appearance, being large and light orange to dark red, with no spots to multiple spots on its back. A primarily tree-dwelling species, the Asian lady beetle quickly abandons the lower ground cover for the tree canopy when aphid populations on the pecan leaves reach an attractive level, making this particular lady beetle a highly effective predator of pecan aphids. Some pecan producers use certain species of crepe myrtle in the orchard to attract Asian lady beetles as well. The crepe myrtle aphid serves as an excellent alternative prey for the lady beetle, since its populations often peak on crepe myrtle plants about two weeks before pecan aphids increase in the pecan trees.[24]

The high degree of plant heterogeneity found in pecan orchards is a boon to diversity and to the management of insect pests. An increase in plant diversity tends to promote an increase in the diversity of animals, from birds to mammals to insects. A varied landscape makes it harder for insects to locate a host plant, leads to a greater diversity of predatory competitors seeking to consume pests, and provides more insects for the predators and parasites to feed on when the pests are not around. This allows the predators and parasites to keep up with the pests. Even so-called weeds can shelter beneficial insects. Without plant diversity in an orchard, pests have little competition and are unleashed to rapidly overcome the crop's defenses, promoting rapid growth in their population.

Pests in a pecan orchard harboring a variety of understory plants or multiple pecan varieties are less likely to find their favorite host plant than when only one type of plant or one pecan variety occupies the orchard. The more often the pests land on a less suitable host plant, the more likely they are to leave the orchard in search of greener pastures.

Insect predators and parasites can specialize in a single pest. An orchard with high plant diversity provides hiding places for some of the pests, which prevents the natural enemies from completely wiping out their prey.[25] Why might this be a good thing? It keeps the predators and parasites in the orchard. On the other hand, those natural enemies that feed on a wide variety of pests have more meal choices in a highly phytodiverse orchard that supports more insect species. Compare it to going to a restaurant with many chefs, each with a different specialty—Chinese food, American, Mexican, seafood, and so forth—versus going to a restaurant that serves only chicken. If you weren't a picky eater, you'd likely spend more time deciding what you wanted to eat, or even sampling a variety of foods in the restaurant with multiple chefs. In another scenario, most adult parasitic wasps eat nothing but the sugar in nectar and pollen, while their young do the work of consuming aphids and other pests. The more plant species in the orchard, the more food sources there are for these insects, which increases the abundance of parasitic wasps that can keep aphids, caterpillars, and other pests in check. Thus, plant diversity tips the scales for the plant species needing protection from predators.

It's not only the species heterogeneity of plants in the orchard that helps keep pest populations down. The presence of trees of varying ages also contributes to a greater diversity of insects, which helps minimize pest numbers in the same ways.[26] Tree-dwelling pests and natural enemies are not the only organisms that benefit in an orchard full of various plant species and trees of varying age and development. Soil-dwelling invertebrates are also more abundant in these situations of structural diversity and complex plant architecture.

Trees are larger, have a more complex architecture, and live longer than herbaceous plants. The varying strata created by a tree's architecture create many habitats and resources for arthropods and other animals.[27] The typical tree provides a predictable source of mating sites, plenty of options for egg-laying locations, refuges in which to hide or rest, and a greater microclimatic range for a host of species. Among the trees' leafy canopies in the glistening sun, shaded undersides of leaves, branches, and miniature canyons of fissures slicing through the bark create niche locations for sun- and shade-loving insects, birds, and other life. As the intensity of the sun changes with its track through the day, the orchard's unseen creatures—beetles, spiders, bark lice, and others—go about their daily business. Tree canopies and thick ground covers alter the reflectivity, temperature, and evapotranspiration of the soil surface and the orchard itself. As ecologist Dan Janzen has suggested, from the perspective of an insect, a tree can serve as a comfortable island in a sea of other vegetation.[28] Also, because trees are larger than islands of annual crops, they can house a greater number of insects based on biomass alone.[29]

All of this may sound quite complex, but in nature, complexity is golden.

Every combination of plants in an orchard may create unique situations and conditions for other species. Usually, the greater the diversity of plant species, the greater the diversity of insects. Why would more bugs be good? Again, complexity. Not all bugs do bad things. Not all bugs in a pecan orchard feed on pecan trees. In fact, plant pest populations are usually lower in more complex systems thanks to the presence of more natural enemies.

Birds are considered reliable indicators of the effect of cultural practices on the environment because of their higher position in the food chain. With the exception of crows and blue jays in North America and parrots in other parts of the world, birds present no harm to the pecan crop and many species use pecan orchards in their daily lives, often doing good from our perspective by feeding on insects. Bird diversity in pecan orchards can be quite surprising. In fact, pecan orchards are often listed among bird-watchers as birding hotspots.[30] For example, an orchard in Bulloch County, Georgia, is noted for the appearance of Baltimore orioles and the rarer Bullock's oriole each winter.[31] A Florida study documented 21 species of birds in a small pecan orchard, including northern mockingbirds, common ground doves, eastern bluebirds, chimney swifts, and purple martins.[32] Orchards are a haven for such birds because of the abundance of insects in the orchard and the cover found there. An Arizona survey listed 66 bird species occurring in a large pecan orchard near Tucson.[33] In my own survey, I documented 47 bird species in orchards across southern Georgia, including unexpected sightings of the swallow-tailed kite in two different orchards. These remarkable fork-tailed raptors feed on reptiles and insects. As I observed them in the orchards, they swooped between tree rows and over treetops feeding on insects and alighting in the tree canopy.

For many birds, most notably the early-successional songbirds, pecan orchards with their diverse architecture, vegetation, and food options are a highly valuable habitat of the Southern agricultural landscape. Where sod and legumes are used as orchard floor cover, birds benefit from an increase in prey and nesting sites. In fall and winter, the branches, trunk, and bark of pecan trees and their innumerable crevices full of insects provide a dining cornucopia and shelter for migrant and wintering birds.

In *Silent Spring*, Rachel Carson wrote, "Over increasingly large areas of the United States, spring now comes unheralded by the return of the birds, and the early mornings are strangely silent where once they were filled with the beauty of birdsong."[34] At one time, this may have been true of commercial pecan orchards. But now, from the health of their soils to the abundance of life forms found within their confines, pecan orchards are full of the indications of sustainability. Among these, birdsong fills the orchards, heralding the resilience found among the pecan, America's favorite tree of life.

· 7 ·

Rebirth

The best time to plant a tree was 20 years ago. The next best time is now.

—Chinese proverb

THE PECAN INDUSTRY HAS BEEN redefined in the twenty-first century by different yet converging discoveries on two sides of the globe. In the Western Hemisphere, where people have been familiar with the pecan for centuries, new discoveries regarding its role in human health have been made. In the East, a burgeoning economy has brought about an awakening to the elements of Western culture—new fads, luxuries, and electronic products—which have made their way from the West to the East in a reversal of the exchange in trade goods that brought such riches as Asian silks, porcelain, and spices to the West centuries before. In this latest exchange, the pecan has become something of a symbol of Western life.

Around 2009, the price of their commodity suddenly began to get better for pecan growers. They could farm pecans at a profit. For years the cost of production had been rising with little change in the price received by those who grew the crop. In 2002, the average cost of producing pecans in the United States was about $850 per acre. By 2008, with skyrocketing fuel prices, the cost had risen to more than $1,500 per acre. During this same period the price received by pecan producers for their crop rose by only 27 cents per pound.[1] At such an exchange, it was costing pecan farmers money just to grow their crop, and in contrast to the case with many other crops, there were no government subsidies to make up the shortfall. The US pecan industry was reeling. But in 2009, China bought nearly one-quarter of the entire US pecan crop, over 80 million pounds worth. Only five years earlier, the United States had exported a mere two million pounds of pecans to China. This phenomenon was fueled by the rapidly growing Chinese middle class, who do not blink at spending nearly $6 for a nine-ounce bag of US-grown pecans, a price nearly six times the minimum wage in Beijing.[2] As a result, the price of pecans for US farmers rose to $1.43 per pound. By 2011, pecans had set record high prices for three years in a row, reaching $2.43 per pound on average. Some farmers received over $3.50 per pound for early harvests of certain varieties.

The similarities between the export of pecans to China in the twenty-first century and the historical exchanges between China and the West are striking. For one, when Chinese trade with the West opened up in the late sixteenth century, China was the wealthiest nation on earth, much as it is today. To the Chinese, the West held little commercial interest. Yet the West had one thing the Chinese coveted: silver.

Today, the United States imports approximately $400 billion worth of products from China each year. By comparison, we export products worth about one-quarter of that amount to China. The Chinese need very little from us with the exception of agricultural products and raw materials for energy and infrastructure. From another perspective, in sixteenth-century China, the arrival of agricultural products like maize and sweet potatoes imported from the West became, in the words of Chinese agricultural historian Song Junling, "one of the most revolutionary events in the history of imperial China."[3] While not exactly a revolutionary event for China, the arrival of pecans introduced the Chinese people to a product for which they developed a seemingly insatiable appetite, much like the one they had once developed for Western silver. This demand in turn helped spark a global revolution in the popularity and production of pecans.

But why did China suddenly go nuts over pecans? Some speculate that it was the bumper crop of pecans produced in the United States in 2007 amid a short walnut crop that introduced the pecan to China, en masse. That year, the price of pecans fell below that of walnuts, something described in the nut business as akin to the price of gold falling below that of silver. This led to the

Chinese purchase of 47 million pounds of US pecans in 2007, a figure over three times that of the previous year.[4] This event was timed with aggressive marketing campaigns in China by a number of pecan-related organizations, including the Georgia Pecan Growers Association, the Texas Pecan Growers Association, the Western Pecan Growers Association, and the National Pecan Shellers Association. As the stars aligned, the Chinese developed a ravenous taste for this uniquely American nut.

The pecan had actually come to China years before, yet the nut had languished in obscurity until the population grew to 1.3 billion people. The pecan was first introduced to China in the early twentieth century when a member of the South American Christian Father's Mission brought along a few nuts for planting in private yards in Jiangyin County in Jiangsu Province, along the Yangtze River about two hours north of Shanghai. From these seeds, 10 trees were produced. Later, in 1907, American botanist E. H. Wilson, who preferred the moniker "Chinese Wilson," took several two-year-old pecan seedlings into China for planting in gardens around Nanjing. Wilson would eventually become renowned in horticultural circles for his introduction of over 1,000 different plants from China into England and the United States, which are today well established in the horticultural trade.[5]

More pecans found their way into China in 1916 and again in 1928, when scientists from Nanking University introduced pecan seeds from Georgia, South Carolina, and Florida, from which a few trees were grown. The first true cultivars were reportedly introduced into China in 1944 by Dr. Fu Huang-guang, although little information regarding the details of these introductions is available. Other early introductions of pecan were made in Putian County, Fujian Province; and in Hangzhou, Jiaxing, and Shaoxing Counties in Zhejiang Province.

An unnamed amateur French horticulturist reportedly sent five grafted Mahan pecan trees to Fujian Province in 1971. Scions from these trees were later distributed to Pingyang County in Zhejiang Province. In 1978, Zhejiang Agricultural College received 14 varieties from the US Pecan Field Station in Brownwood, Texas. Through all of these and other introductions, many pecan seeds and varieties have been introduced into China over the last 100 years or more. Although many experimental and small commercial orchards were established during this time, primarily in Jiangsu and Zhejiang Provinces, pecans also occurred in small plantings or as specimen trees from about 24° to 40° N, an area extending from Fujian Province to Beijing.[6]

During the 1980s more than 100,000 pecan trees were reported in Nanjing and Hangzhou. At the same time, a total of 15,000 ornamental plantings of pecan were found in Nanjing. At one time, the Nanjing Botanical Garden itself had more than 1,000 pecan trees. Much of the material introduced into China early on was in the form of seeds or seedling trees, so there is a moderately high diversity of pecan in China. This population provides a large base of genetic diversity for the selection of trees adapted to the growing conditions found there. Chinese horticulturists have been at work selecting and breeding pecan cultivars since the 1950s. In the late 1990s, scientists from southern China requested pecan propagation materials from scientists at the US Pecan Field Station. Shipments of graftwood and seedstock were made to Hunan, Yunnan, and Anhui Provinces in 1999 and 2000.[7]

While China certainly has the potential to commercially produce pecans, its per capita production may be limited to some extent by the availability of suitable land for intensive pecan production. Pecans grow best on flat or gently sloping land with deep, fertile soils. In China, much of the land that would be best suited to pecan production is used to produce staple foods like rice, corn, and soybeans to feed the burgeoning population. Much of the remaining land available to pecan production occurs along steep slopes and hillsides. Studies by Chinese horticulturists have substantiated this observation by showing that pecan trees grow much better on their soils in areas with flat topography as opposed to thin soils along the dry, steep slopes of the hillsides. Still, the Chinese are renowned for their ingenuity and have developed horticultural methods to suit their needs for the production of many fruits and nuts throughout history. New pecan orchards have been planted in China over the last decade as far north as Yunnan Province (24° N latitude) to as far south as Henan Province (34° N latitude), corresponding to a range equivalent to the distance from

Kansas to Durango and Coahuila, Mexico. Research programs are currently underway in China to study the adaptability of the species to the area.[8]

I went to China in 2010 at the request of the Georgia Pecan Growers Association in order to attend a large food show in Shanghai. We offered samples of in-shell and shelled pecans wrapped in nice little cellophane bags. From the start, it was clear that many of the Chinese who came by our booth had never seen a pecan before. We showed them how to shell the nuts. They sniffed them, turned them over in their hands, and finally tasted them. Soon, we had one of the most popular booths at the show. As a side trip prior to the show, our group saw an orchard of Chinese hickory trees, *Carya cathayensis*, growing on the side of a steep hill near a small village west of Hangzhou. The trees were marked with numbers and Chinese lettering. Spikes were driven into the sides of the trees for workers to climb on to knock the nuts out at harvest. One of my colleagues took the initiative to climb a tree, to which our hosts strongly objected, showing their disapproval by waving their arms and admonishing him to get down out of the tree. We were shown the fruit of these trees, which were small nuts about the diameter of a nickel. They were hard to crack and contained very little kernel. Eating one is a lot of work for little reward, yet they are a traditional snack in this area of China. Upon introduction, Chinese consumers suddenly found that pecans tasted good and were easier to shell than the native Chinese hickory and other nuts. They had stumbled on a revelation: pecans were more than just a funny-looking walnut.

Sustainable demand for pecans in the future will require the development of international markets beyond China alone. Efforts are underway to establish the pecan as a viable food commodity in India, Turkey, and other areas of the Middle East because of the long-standing proclivity for tree nuts in these regions. Europe, Canada, South Korea, and Russia have also been targeted as marketing points to help expand the demand for and renown of the pecan. Another strong advantage for pecans is that the world is becoming more health conscious. At about the same time that the international push to market pecans began, new research in the United States was showing that the pecan was of significant health value to humans.

The Tree of Life

Mediterranean diets based largely on fruits, vegetables, fish, whole grains, legumes, herbs, spices, seeds, nuts, and olive oil have long been known to contribute to human health. The dietary habits of those in the Mediterranean region have been linked to a lower incidence of coronary heart disease and other chronic illnesses.[9] Although Mediterranean diets are rich in fats, they are based on so-called healthy fats derived from olive oil and nuts.

The pecan kernel contains high levels of oil and fatty acids. The percentage of oil found in a pecan kernel can range from less than 60 percent to more than 75 percent, depending on its genetic makeup and growing conditions.[10] As the pecan crop develops, the oil content of the nut increases rapidly as it approaches maturity. In fact, the kernel's entire dry weight accumulates during a six-week period beginning when the nuts have reached their full size. The lipid content of pecans is heavily influenced by irregular bearing. While there is much variation between pecan varieties, in general, trees bearing a heavy nut crop have lower lipid levels than those bearing a light crop.

The pecan has taken a bad rap over the years as a result of the sugary, tasty pies and pralines by which most people have come to know this nut. In fact, the pecan nut itself is one of the healthiest foods you can eat. Once upon a time, all fats were considered bad and were thought to be responsible for a wide variety of diseases, including cardiovascular disease and diabetes. We now understand that not all fats are created equal. There are good fats and bad fats and everything in between. Contrary to popular belief, you shouldn't fear gaining weight from eating nuts like pecans, which may have a higher fat content than some foods but are whole foods that are high in fiber. As a result, the fat found in pecans and other nuts is less absorbable than the fats found in processed foods. In fact, people who consume three to five servings of nuts per week maintain a healthy body weight better than those who do not eat nuts.[11]

The so-called good fats are unsaturated fats, and they help protect us from the very health problems that consuming excess saturated fat was believed to cause. Over 90 percent of the fat found in pecans is unsaturated fat, which can be divided into monounsaturated fats and polyunsaturated fats, both of which have great benefit for cholesterol levels. Monounsaturated fats help lower bad cholesterol, otherwise known as low-density lipoproteins (LDL), and raise good cholesterol, or high-density lipoproteins (HDL). Polyunsaturated fats help lower total cholesterol, including HDL, but they are rich in the omega-3 fatty acids, which have been shown to lower blood pressure, combat LDL, fight inflammation, and protect the nervous system. One omega-3 fatty acid is an essential fatty acid and cannot be manufactured by our bodies. Therefore, it must be obtained from the foods we consume, and one rich source of it is nuts, including the pecan.[12]

Perhaps the first real boost to the pecan's image from a human nutrition standpoint came from a study published in 2001 by a group of scientists from Loma Linda University in California. The study, led by Dr. Sujatha Rajaram, demonstrated that substituting 20 percent of the calories from the American Heart Association's Step 1 diet with pecans lowered total cholesterol by 11.5 percent, while raising good cholesterol or HDL without causing weight gain.[13] Multiple studies since that time have confirmed the positive effects of pecans on the lowering of bad cholesterol or LDL.

Our bodies are constantly at war with infection and disease. The normal day-to-day interaction of our bodies with our environment, such as the air we breathe or the foods we eat along with our lifestyle habits like smoking or the lack of exercise, cause the development of substances called free radicals that attack healthy cells. When healthy cells are weakened, our bodies become more susceptible to health problems like heart disease and certain types of cancers. Antioxidants are substances that help protect healthy cells from damage caused by free radicals.

The pecan is ranked among the foods with the highest phenolic content, a measure of phenolic acid, which is a very potent antioxidant. In a report that screened common foods across the United States, pecan kernels were shown to have the highest antioxidant capacity and total extractable phenolic content of all nuts.[14] Surprisingly, pecans were ranked even higher than many fruits, such as blueberries, which are commonly marketed for their high antioxidant properties. Various phytochemicals found in the pecan contribute to the nut's high antioxidant and radical-scavenging capacities. Phenolic acids, flavonols, proanthocyanidins, tannins, and tocopherols account for the majority of these secondary plant metabolites of pecan. Many of these compounds are the same ones responsible for the pecan's distinctive taste.[15]

Further studies by Dr. Rajaram revealed that a pecan-enriched diet significantly raised blood levels of gamma tocopherol as a result of the high amounts of naturally occurring gamma tocopherol found in pecans. Tocopherols act as a natural defense against oxidative events that can lead to off flavor and darkening of the seed coat. For humans, this unique form of vitamin E is an important antioxidant that may benefit intestinal health and have a protective effect against prostate cancer and other chronic diseases.

Still other studies have shown that a diet rich in pecans increases levels of dietary fiber, thiamine, magnesium, copper, and manganese. Pecans are also notably high in protein. Proteins are composed of amino acids, one of which, arginine, is found at high levels in pecan and helps relax blood vessels and maintain healthy blood pressure.[16] The many nutritional qualities of the pecan have drawn the attention of health organizations like the National Heart Association, which has greatly contributed to the nut's recent popularity around the world. Epidemiological studies have consistently demonstrated that nut consumption reduces the risk of coronary heart disease by helping to lower blood cholesterol, maintain blood pressure, and sustain normal blood flow to the tissues, thus preserving the health of blood vessel walls and slowing atherosclerosis or hardening of the arteries.[17] The US Food and Drug Administration has even approved a qualified health claim stating, "Scientific evidence suggests, but does not prove, that eating 1.5 ounces per day of some nuts, as part of a diet low in saturated fat and cholesterol, may reduce the risk of heart disease."

The most significant aspect of the pecan's cardiovascular health benefit is its blood lipid–lowering characteristics. Still, this may only partially explain the risk reduction. The pecan is low in saturated fatty acids and rich in monounsaturated fatty acids, especially oleic acid, which has a positive effect on blood lipids. These nutritional attributes indicate that the pecan can serve as an important healthy food in the human diet. In fact, it is now known that eating as little as one ounce of nuts per day is associated with a prolonged life by reducing the risk of multiple diseases, including type 2 diabetes mellitus, colon cancer, hypertension, gallstone disease, diverticulitis, and inflammatory diseases.[18] As we learn more about the contributions of pecans to a healthy lifestyle and the human population becomes more health conscious, there will be a greater demand for foods like pecans. This increase in demand will, of course, require a greater and more consistent supply of pecans.

Not surprisingly, the strong interest and wildly improved prices have generated a planting boom for the pecan not seen since the 1920s. In Georgia, commercial pecan acreage hovered around 140,000 acres for 20 years or more. Prior to this boom, as quickly as orchards were planted, many were lost to development or abandoned. There were many years in which the pecan acreage even declined. This downward trend changed with the opening of China as an additional market for the crop. As prices became profitable, new orchards began popping up at a remarkable pace.

In 2010, at least 68,000 pecan trees were planted in Georgia. About half of these were planted as replacement or interplanted trees in existing orchards. The other half accounted for at least 1,300 acres of new pecan orchards, a relatively modest number but still up from the usual planting trend. In addition, nearly 900 acres of abandoned orchards were brought back into production in an effort to quickly increase yields.[19] The pecan boom was beginning, but it was not until the winter of 2012 that things really began to grow on a grand scale. Pecan nurseries throughout the Southeast sold out quickly and advance sales for the next two years were rapidly gobbled up. Existing pecan nurseries began expanding their stock and new nurseries began to spring up across the region.

Following two years of record prices, well over 100,000 pecan trees were planted in the state in 2012, 87 percent of which went into the establishment of new orchards. The remaining trees were planted to replace lost trees or were planted in open spaces within existing orchards. Desirable and Pawnee accounted for 43 percent of pecan trees planted. Cape Fear, Sumner,

Excel, Oconee, and Caddo were also popular varieties. In all, Georgia producers commonly planted at least 18 different varieties that year. Conspicuous by its absence was Stuart. The most popular pecan tree planted throughout the Southeast since the initial establishment of improved cultivar orchards at the dawn of the twentieth century, Stuart was not planted in significant numbers and failed to show up on surveys of planted trees.[20]

The most conservative estimate places the 2012 newly planted pecan orchard acreage in Georgia at about 3,800 acres; nearly three times the area planted two years before. However, this is a very conservative estimate based on a limited survey.[21] Southeastern pecan nursery sales in 2012 suggest that 235,500 trees were sold that year, the majority of which went to Georgia orchards. These data would suggest that the newly planted acreage in the state was closer to 5,000 to 6,000 acres rather than 3,800. Add to this the nearly 1,800 acres of abandoned trees being brought back into production, and the number escalates rapidly.

Although pecan acreage in Georgia is rising, it's not certain that all of these trees and orchards will contribute to the state's pecan production in the next 10 years. While many of those growers planting pecan trees are experienced farmers, many others are new to pecan farming. Some of those planting trees may be unwilling or unable to provide the trees with everything needed to develop into productive orchards. I often suggest to those interested in growing pecans or planting an orchard for the first time, "Don't do it unless you will enjoy it—all of it—the sweat, toil, skinned knuckles, and sore back as well as the bounty." If you are planning on growing pecans to get rich, choose another endeavor.

Yet Georgia's pecan production will likely rise with the increase in acreage. The same trend is occurring across the pecan belt, although not at the scale it is occurring in Georgia. In addition, pecan-producing countries throughout the world are renewing interest in the crop and planting additional land to pecan trees. With the marketing of pecans in the United States and abroad only just beginning, these trees will be needed to meet the world's growing demand for the nut.

While the development of the pecan as a crop has come far, there remain untold chapters to be written in the crop's rich story. Chinese demand has helped fuel a rebirth in the American pecan industry, causing an economic bubble around the pecan early in the twenty-first century. Like all bubbles, though, the pecan bubble will likely burst if it continues its rapid growth without an effective increase in consumption. Further expansion of export markets is required, as well as a fresh, large-scale approach to domestic marketing in the United States, both of which need long-term cooperation between all segments and regions of the pecan industry to help sustain the growth and life of this crop.

So, where is the pecan headed today? The industry that has developed from this remarkable nut is expanding worldwide, changing the global dynamics of the pecan as a crop. The United States, by far the world's largest producer of pecans, is at a crossroads. As I write this in 2014, the pecan planting boom that began at the beginning of this decade continues, fueled by the fumes of record high prices obtained for the 2010 and 2011 crops, thanks to the Asian export market. While export prices for pecans have tempered somewhat, they remain profitable. The US domestic market for pecans, on the other hand, has remained stagnant, stalling the value of the nation's pecan crop, while the value of the other tree nuts—almonds, pistachios, and walnuts—has grown by an average of 98 percent since 2002 in the face of acreage expansions averaging 194 percent. There remains plenty of room for growth of the US pecan market. The pieces are now in place to better establish the pecan as a regular part of the human diet in its native home and throughout the world. New advances are on the horizon, including additional revelations about the nutritional benefits of pecan. Scientists are tapping into the pecan tree's genetic code. New techniques are being developed to further enhance the efficiency and sustainability of pecan cropping systems. Pecan scientists and producers continue to learn and adapt methods of production to the tree's needs in an attempt to sustainably optimize yields and production efficiency. Relatively consistent pecan yields have recently been achieved with progressive management practices that manipulate and utilize the tree's own biological characteristics. Pecans are being marketed around the world and the previously disjointed pecan industry is more unified than ever before.

The nut that was utilized for millennia by Native Americans, discovered by the early explorers of North America, and developed over the course of US history into a valuable cash crop is still evolving and moving forward. America's tree of life continues its journey alongside its most significant partner, humankind.

Pecan Recipes

Toasted Pecans

2 cups pecan halves
2 teaspoons butter
½ teaspoon sea salt

Place pecans in a single layer on a large baking sheet in a 275°F oven. After 20 minutes, add the butter and stir pecans until they are thoroughly coated. Sprinkle with ½ teaspoon of salt. Replace in oven and repeat process every 15 minutes, without adding salt unless a taste test dictates more. Allow pecans to roast for at least an hour. Serve warm or cooled.

Pecan Waffles

2 cups sifted flour
3 teaspoons baking powder
1 teaspoon salt
¾ cup chopped pecans
2 eggs
1¼ cups whole milk
½ cup melted butter

Preheat waffle iron according to manufacturer's instruction. Sift together flour, baking powder, and salt. Stir in pecans. Make a well in the center of the dry ingredients. Beat eggs slightly and add milk. Pour egg and milk mixture into well and mix. Add melted butter and stir. Pour about ½ cup of the batter onto hot iron for a single waffle. Close iron. Cook until brown and crisp.

Lemon Broccoli with Pecans

½ to 1 cup chopped pecans
1 tablespoon butter
1½ pounds fresh broccoli
¼ cup lemon juice
2 teaspoons cornstarch
½ cup diluted canned chicken broth
1 tablespoon grated lemon rind
1 tablespoon granulated sugar
¼ teaspoon pepper
salt to taste

Sauté the pecans in the butter in a skillet until lightly browned. Set aside. Wash the broccoli and separate it into spears. Steam in a steamer until tender-crisp, then drain. Transfer to a serving bowl, cover, and keep warm. Mix lemon juice and cornstarch in a saucepan. Stir in the chicken broth, lemon peel, sugar, pepper, and salt. Cook over medium heat until mixture is thickened, stirring constantly. Pour the lemon sauce over the broccoli. Sprinkle with the pecans and serve immediately.

(Courtesy of the Junior League of Cobb-Marietta, Inc., in *Southern on Occasion: A Companion to Inspire Gracious Living*, 1998)

Sweet Potato Casserole

4 cups cooked sweet potatoes, mashed
½ cup sugar
2 eggs
1 teaspoon vanilla extract
½ cup butter
⅓ cup milk
Topping:
1 cup light brown sugar
1 cup chopped pecans
⅓ cup flour
⅓ cup butter

Mash sweet potatoes in an electric mixer. Add sugar, eggs, vanilla, butter, and milk and mix together. Ladle sweet potato mixture into a greased casserole dish. Combine brown sugar, chopped pecans, flour, and butter in a bowl and sprinkle on top of potatoes. Bake at 350° F for 25–30 minutes.

Pecan-Encrusted Fish

½ cup pecan pieces
½ cup bread crumbs
1 pound fresh fish fillets (grouper / snapper / trout / catfish)
salt and pepper to taste
⅓ cup flour
2 eggs, beaten
¾ cup butter
1 lemon, juiced
1 bunch of fresh parsley, chopped

Process the pecans and bread crumbs in a food processor just until a coarse mixture forms. Season the fillet pieces with salt and pepper, dredge in the flour, dip in the egg mixture, and coat with the pecan mixture.

Melt ¼ cup butter in a nonstick ovenproof skillet over medium heat. Sauté the fish on one side until it is brown. Turn it and sauté the other side of the fish. Bake at 400°F for 10 minutes or until the fillet pieces flake easily. Remove the fish to a warm platter; wipe the skillet.

Add the remaining butter to the skillet. Cook over high heat until the butter is foamy and dark brown, stirring constantly. Add the lemon juice and parsley, still stirring. Pour over the fillets. Serve immediately.

Sunday Chicken Salad

2 whole (or 4 halves) chicken breasts
¾ cup chopped green celery
1 cup halved seedless grapes (purple or green)
¾ cup pecan pieces
dash of salt and pepper to taste
1 cup mayonnaise

Simmer chicken breasts in slightly salted water to cover until tender. Allow to cool and cut rough pieces into a large mixing bowl. Add the celery, grapes, and pecan pieces and mix thoroughly. Then add the mayonnaise and flavorings and mix until a nice spreadable consistency is reached. Refrigerate in tightly covered container. Salad may be served in lettuce cups; in a whole, quartered small tomato; or between slices of freshly baked asiago-cheese bread.

Congealed Fruit Salad

1 small can applesauce
2 small packages (3 oz.) cherry gelatin
1 cup ginger ale
1 small can crushed pineapple
½ to 1 cup chopped pecans

Heat applesauce to boiling point and remove from heat. Stir gelatin into the applesauce. Add ginger ale and chill until set. Stir in the pineapple and nuts and chill until firm. Slice into squares and serve on bed of lettuce. A dollop of whipped cream on the top is optional.

Pecan-Coated Pork Tenderloin

¼ cup apple cider
1 (1 lb.) pork tenderloin, trimmed
¼ cup packed light brown sugar
1 tablespoon spicy brown mustard
½ teaspoon salt
¼ teaspoon black pepper
2 cloves garlic, minced
⅔ cup finely chopped pecans

Combine apple cider and pork tenderloin in a zip-top freezer bag; seal and marinate in refrigerator for 8 hours. Preheat oven to 400°F. Remove pork from bag; discard cider. Combine sugar, mustard, salt, pepper, and garlic and rub mixture over pork. Roll pork in pecans. Place pork on broiler-pan rack coated with cooking spray. Bake at 400°F for 25 minutes or until a thermometer registers 160°F. Remove from oven and let stand for 10 minutes before slicing and serving.

Southern Pecan Pie

1¼ cups light cane syrup
1 cup granulated sugar
4 fresh eggs
4 tablespoons melted butter
1 teaspoon vanilla
1½ cups pecan pieces

Cook sugar and syrup together for 3 minutes. Beat eggs together and pour mixture slowly into hot syrup, adding the butter, vanilla, and pecan pieces. Pour into an uncooked pie crust and bake in a 350° F oven about 45 minutes, or until set.

Pecan Pralines

1 cup granulated sugar
1 cup brown sugar
2 tablespoons light corn syrup
4 tablespoons butter
½ cup cream
¼ teaspoon cream of tartar
¼ teaspoon salt
2 cups pecan pieces

Mix sugars, syrup, butter, cream, cream of tartar, and salt in a saucepan. Boil to the soft ball stage, between 234°F and 240°F, using a candy thermometer. Remove from heat and beat for 2 minutes. Cool 15 minutes. Beat again 10 to 15 whips. Add pecans and drop onto waxed paper. **NOTE:** If mixture becomes too hard, beat in a small amount of hot water.

Green Gelatin Salad

1 (3-ounce) package green gelatin
1 cup boiling water
8 ounces small marshmallows
1 cup cold water
4 ounces cream cheese
½ cup mayonnaise
1 (8-ounce) package whipped topping
1 (8-ounce) can crushed pineapple in juice, undrained
¼ to ½ cup coarsely chopped pecans

Dissolve gelatin package into boiling water over low heat. Stir in marshmallows just until melted. Remove from heat and stir in cold water; set aside. Cream together cream cheese and mayonnaise; beat until creamy. Stir into gelatin mixture, then stir in whipped topping, pineapple, and chopped pecans. Pour into mold or serving bowl and refrigerate until set, at least 4 hours or over night.

Notes

Introduction

1. "The Kitchen," *Texas Siftings* (Austin, TX), February 6, 1886.

2. S. Murch, *Like Wine and Cheese . . . Older Is Better.* City of Dallas, TX (1974).

3. "Favorite Recipe," *Democrat-American.* (Sallisaw, OK), February 19, 1931.

4. M. K. Rawlings, *Cross Creek Cookery* (New York: Charles Scribner's Sons, 1942).

5. P. Daley, "Glencoe Man Pursues Perfection through Pecan Pie," *Chicago Tribune*, November 20, 1994.

6. A. Davidson, *Oxford Companion to Food* (Oxford: Oxford University Press, 1999).

Chapter 1

1. D. R. Layne and D. Bassi, eds., *The Peach: Botany, Production and Uses* (Oxfordshire, UK: CAB International, 2008).

2. R. H. True, 1917. "Notes on the Early History of the Pecan in America," *Smithsonian Institution Annual Report* (1917): 435–48.

3. J. H. Trumbull, "Words Derived from Indian Languages of North America," *Transactions of the American Philological Association* 4 (1872): 19–32.

4. D. Rowland, *Encyclopedia of Mississippi History: Comprising Sketches of Counties, Towns, Events, Institutions, and Persons* (Madison, WI: Selwyn A. Brant, 1907).

5. A. Penicaut, *Fleur de Lys and Calumet: Being the Penicaut Narrative of French Adventure in Louisiana*, trans. R. G. McWilliams (Baton Rouge: Louisiana State University Press, 1953).

6. J. R. Flack, "The Spread and Domestication of the Pecan in the U.S." (PhD diss., University of Wisconsin, 1970).

7. H. Marshall, *Arbustrum Americanum: The American Grove* (Philadelphia: Joseph Cruikshank, 1785).

8. L. J. Grauke, J. W. Pratt, W. J. Mahler, and A. O. Ajayi, "Proposal to Conserve the Name of Pecan as *Carya illinoensis* and Reject the Orthographic Variant *Carya illinoinensis*," *Taxon* 35 (1986): 174–77.

9. L. J. Grauke, "Appropriate Name for Pecan," *HortScience* 26 (1991): 1358.

10. P. S. Manos and D. E. Stone, "Evolution, Phylogeny, and Systematics of the Juglandaceae," *Annals of the Missouri Botanical Garden* 88 (2001): 231–69.

11. S. R. Manchester, "The Fossil History of the Juglandaceae," *Monographs in Systematic Botany from the Missouri Botanical Garden* 21 (1987): 1–137.

12. H. Lee, *To Kill a Mockingbird* (Philadelphia: J. B. Lippincott, 1960).

13. J. R. Flack, "The Spread and Domestication of the Pecan in the U.S." (PhD diss., University of Wisconsin, 1970).

14. Ibid.

15. R. M. Harper, *Economic Botany of Alabama. Part 2. Catalogue of the Trees, Shrubs and Vines of Alabama, with Their Economic Properties and Local Distribution* (Geological Survey of Alabama, University of Alabama, 1928).

16. P. D. Welch and C. M. Scarry, "Status-Related Variation in Foodways in the Moundville Chiefdom," *American Antiquity* 60 (1995): 397–419.

17. D. R. Whitehead, "Pollen Morphology in the Juglandaceae, I: Pollen Size and Pore Number Variation," *Journal of the Arnold Arboretum* 44 (1963): 101–10.

18. L. J. Grauke, M. Azucena Mendoza-Herrera, A. J. Miller, and B. W. Wood, "Geographic Patterns of Genetic Variation in Native Pecans," *Tree Genetics and Genomes* 7 (2011): 917–32.

19. Ibid.

20. Ibid.

21. E. A. Bettis, R. G. Baker, B. K. Nations, and D. W. Benn, "Early Holocene Pecan, *Carya illinoensis*, in the Mississippi River Valley near Muscatine, Iowa," *Quaternary Research* 33 (1990): 102–7.

22. M. R. Gilmore, *Uses of Plants by the Indians of the Missouri River Region*, Smithsonian Institution Bureau of American Ethnology Report 33 (1919).

23. J. R. Flack, "The Spread and Domestication of the Pecan in the U.S." (PhD diss., University of Wisconsin, 1970).

24. E. A. Bettis, R. G. Baker, B. K. Nations, and D. W. Benn, "Early Holocene Pecan, *Carya illinoensis*, in the Mississippi River Valley near Muscatine, Iowa," *Quaternary Research* 33 (1990): 102–7.

25. M. L. Fowler, "Modoc Rock Shelter an Early Archaic Site in Southern Illinois," *American Antiquity* 24 (1959): 257–70.

26. A. Resendez, *A Land So Strange: The Epic Journey of Cabeza de Vaca* (New York: Basic Books, 2007).

27. E. H. Johnson, "Edwards Plateau," Handbook of Texas Online, accessed May 3, 2010, https://tshaonline.org/handbook/online/articles/rxe01.

28. T. R. Hester, "Late Paleo-Indian Occupations at Baker Cave, Southwestern

Texas," *Bulletin of the Texas Archeological Society* 53 (1981): 101–19.

29. J. M. Quigg, J. D. Owens, G. D. Smith, and M. Cody, *The Varga Site: A Multi-component, Stratified Campsite in the Canyonlands of Edwards County, Texas*, Technical Report No. 35319 (Austin, TX: TRC, 2005).

30. J. R. Flack, "The Spread and Domestication of the Pecan in the U.S." (PhD diss., University of Wisconsin, 1970).

31. G. Dorsey, *Traditions of the Caddo* (Lincoln: University of Nebraska Press, 1997).

32. G. D. Hall, "Pecan Food Potential in Prehistoric North America," *Economic Botany* 54 (2000): 103–12.

33. J. R. Flack, "The Spread and Domestication of the Pecan in the U.S." (PhD diss., University of Wisconsin, 1970).

34. G. D. Hall, "Pecan Food Potential in Prehistoric North America," *Economic Botany* 54 (2000): 103–12.

35. J. R. Flack, "The Spread and Domestication of the Pecan in the U.S." (PhD diss., University of Wisconsin, 1970).

36. Ibid.

37. V. M. Rose, "The Life and Services of General Ben McCulloch," *The Steck Company* (Austin, Texas, 1958).

38. G. P. Winship, ed. and trans., *The Journey of Coronado, 1540–1542, from the City of Mexico to the Grand Canyon of the Colorado and the Buffalo Plains of Texas, Kansas, and Nebraska, as Told by Himself and His Followers* (New York: A. S. Barnes, 1904).

39. J. R. Flack, "The Spread and Domestication of the Pecan in the U.S." (PhD diss., University of Wisconsin, 1970).

40. R. G. Thwaites, ed. *The Jesuit Relations and Allied Documents* (Cleveland: Burrows, 1986).

41. J. R. Flack, "The Spread and Domestication of the Pecan in the U.S." (PhD diss., University of Wisconsin, 1970).

42. H. E. Bolton, *Original Narratives of Early American History: Spanish Exploration in the Southwest, 1542–1706* (New York: Scribner, 1916).

43. E. Berkeley and D. S. Berkeley, *The Life and Travels of John Bartram: From Lake Ontario to the River St. John* (Gainesville: University of Florida Press, 2009).

44. W. Bartram, *Travels of William Bartram*, ed. Francis Harper (New Haven, CT: Yale University Press, 1958).

45. G. Onderdonk, "Pomological Possibilities of Texas," *Texas Department of Agriculture Bulletin* 9 (1911): 18.

46. J. R. Flack, "The Spread and Domestication of the Pecan in the U.S." (PhD diss., University of Wisconsin, 1970).

47. D. Fairchild, *Exploring for Plants* (New York: Macmillan, 1930).

48. P. Cornett, "Encounters with America's Premier Nursery and Botanic Garden." *Twinleaf,* January 2004, accessed August 16, 2010, http://www.monticello.org/sites/default/files/inline-pdfs/2004_encounters.pdf.

49. W. P. Corsa, *Nut Culture in the United States* (Washington, DC: US Department of Agriculture, 1896).

50. A. Fusonie, *George Washington: Pioneer Farmer* (Charlottesville: University of Virginia Press, 1998).

51. J. R. Flack, "The Spread and Domestication of the Pecan in the U.S." (PhD diss., University of Wisconsin, 1970).

52. R. H. True, 1917. "Notes on the Early History of the Pecan in America," *Smithsonian Institution Annual Report* (1917):435–48.

53. T. Jefferson, *Notes on the State of Virginia* (New York: W. W. Norton, 1982).

54. J. R. Flack, "The Spread and Domestication of the Pecan in the U.S." (PhD diss., University of Wisconsin, 1970).

55. Ibid.

56. H. Savage and E. Savage, *Andre and Francois Andre Michaux* (Charlottesville: University of Virginia Press, 1986).

57. W. S. Bryant, "Botanical Explorations of Andre Michaux in Kentucky: Observations of Vegetation in the 1790s," *Castanea Occasional Papers* 2 (2004): 211–16.

CHAPTER 2

1. M. Twain, *Life on the Mississippi* (New York: Signet Classics, 1961).

2. Ibid.

3. E. M. Summers, E. W. Brandes, and R. D. Sands, *Mosaic of Sugarcane in the United States, with Special Reference to Strains of the Virus,* Technical Bulletin 955 (Washington, DC: US Department of Agriculture, 1948).

4. L. H. Bailey, *The Nursery-Book, a Complete Guide to the Multiplication of Plants* (New York: Macmillan, 1914).

5. K. Mudge, J. Janick, S. Scofield, and E. E. Goldschmidt, "A History of Grafting," *Horticultural Reviews* 35 (2009): 437–93.

6. Ibid.

7. L. J. Grauke and T. E. Thompson, "Rootstock Development in Temperate Nut Crops," *Acta Horticulturae* 622 (2003): 553–66.

8. J. B. Koverman, "Clay Connections: A Thousand Mile Journey from South Carolina to Texas," in *American Material Culture and the Texas Experience: The David B. Warren Symposium* (Bayou Bend Collection and Gardens at the Museum of Fine Arts, Houston, Texas, 2009). Accessed August 21, 2012. http://scholarcommons.sc.edu/cgi/viewcontent.cgi?article=1022&context=mks_staffpub.

9. L. Todd, *Carolina Clay: The Life and Legend of the Slave Potter Dave* (New York: Norton, 2008).

10. A. Landrum, letter to the editor, *American Farmer* 4 (1822): 7.

11. "Slavery at Oak Alley Plantation," accessed June 14, 2010, http://www.oakalley-plantation.com/SlaveryResearchDatabase

12. J. R. Flack, "The Spread and Domestication of the Pecan in the U.S." (PhD diss., University of Wisconsin, 1970).

13. W. A. Taylor, *Promising New Fruits*, Yearbook of the Department of Agriculture (Washington, DC: US Department of Agriculture, 1905).

14. Ibid.

15. Anonymous, "Mr. Emil Bourgeois," *Louisiana Planter and Sugar Manufacturer* 30, no. 17 (1903): 259.

16. D. Sparks, *Pecan Cultivars* (Watkinsville, GA: Pecan Production Innovations, 1992).

17. R. R. Wolfe, "Biography of E. E. Risien," *Proceedings, Texas Pecan Growers Association* 25 (1946): 48–53.

18. A. W. Hamrick, *The Call of the San Saba: A History of San Saba County, Texas* (San Antonio, TX: Naylor County, 1941).

19. "San Saba Mother Pecan," Famous Trees of Texas, Texas Forest Service, accessed August 23, 2012, http://texasforestservice.tamu.edu/websites/FamousTreesOfTexas/TreeLayout.aspx?pageid=16138..

20. R. R. Wolfe, "Biography of E. E. Risien," *Proceedings, Texas Pecan Growers Association* 25 (1946): 48–53.

21. T. E. Dabney, *Ocean Springs: The Land Where Dreams Come True* (Pascagoula, MS: Lewis Printing Services, 1974).

22. Anonymous, *The History of Jackson County, Mississippi* (Pascagoula, MS: Jackson Genealogical Society, 1989), 19–20.

23. G. E. Kenknight, "Pecan Varieties Happen in Jackson County, Mississippi," *Pecan Quarterly* 4, no. 3 (1970): 6–7.

24. W. A. Taylor, *Promising New Fruits*, Yearbook of the Department of Agriculture (Washington, DC: US Department of Agriculture, 1905).

25. Anonymous. 1890. "How to Grow Pecans: A Talk with Colonel Stuart of Mississippi," *Atlanta Constitution*, November 2, 1890.

26. R. Bellande, "Pabst Family," Ocean Springs Archives, accessed October 2, 2011, http://www.oceanspringsarchives.net/node/78.

27. J. R. Flack, "The Spread and Domestication of the Pecan in the U.S." (PhD diss., University of Wisconsin, 1970).

28. R. Bellande, "Pabst Family," Ocean Springs Archives, accessed October 2, 2011, http://www.oceanspringsarchives.net/node/78.

29. D. Sparks, *Pecan Cultivars* (Watkinsville, GA: Pecan Production Innovations, 1992).

30. Ibid.

31. Anonymous, "Charles Forkert, Ocean Springs, Miss.," *American Nut Journal* (1921).

32. C. Forkert, "Twelve Years Experience in Hybridizing Pecans," *Proceedings of the National Nut Growers Association* (1914): 28–30.

33. D. Sparks, *Pecan Cultivars* (Watkinsville, GA: Pecan Production Innovations, 1992).

34. M. L. Wells, "Pecan Planting Trends in Georgia," *HortTechnology* 24 (2014): 475–79.

35. "Theo Bechtel Funeral Held This Morning," *Daily Herald* (Biloxi, MS), January 19, 1931.

36. R. Bellande, "Pabst Family," Ocean Springs Archives, accessed October 2, 2011, http://www.oceanspringsarchives.net/node/78.

37. D. Sparks, *Pecan Cultivars* (Watkinsville, GA: Pecan Production Innovations, 1992).

CHAPTER 3

1. D. Sparks, "Adaptability of Pecan as a Species," *HortScience* 40 (2005): 1175–89.

2. M. W. Smith, B. L. Carroll, and B. S. Cheary, "Chilling Requirement for Pecan," *Journal of the American Society for Horticultural Science* 117 (1992): 745–48.

3. B. W. Wood, L. J. Grauke, and J. R. Payne, "Provenance Variation in Pecan," *Journal of the American Society for Horticultural Science* 123 (1998): 1023–28.

4. D. Sparks, "Chilling and Heating Model for Pecan Budbreak," *Journal of the American Society for Horticultural Science* 118 (1993): 29–35.

5. D. Sparks, "Adaptability of Pecan as a Species," *HortScience* 40 (2005): 1175–89.

6. B. W. Wood, L. J. Grauke, and J. R. Payne, "Provenance Variation in Pecan," *Journal of the American Society for Horticultural Science* 123 (1998): 1023–28.

7. F. R. Brison, *Pecan Culture* (College Station: Texas Pecan Growers Association, 1974).

8. J. R. Smith, *Tree Crops: A Permanent Agriculture* (Old Greenwich, CT: Devin-Adair, 1977).

9. M. Sagaram, L. Lombardini, and L. J. Grauke, "Variation of Leaf Anatomy of Pecan Cultivars from Three Ecogeographic Locations," *Journal of the American Society for Horticultural Science* 132 (2007): 592–96.

10. D. Sparks, "Adaptability of Pecan as a Species," *HortScience* 40 (2005): 1175–89.

11. P. C. Andersen, and B. V. Broadbeck, "Net CO_2 Assimilation and Plant Water Relations Characteristics of Pecan Growth Flushes," *Journal of the American Society for Horticultural Science* 113 (1988): 444–50.

12. D. Sparks, "Adaptability of Pecan as a Species," *HortScience* 40 (2005): 1175–89.

13. Ibid.

14. J. A. Putnam and H. Bull, *The Trees of the Bottomlands of the Mississippi River Delta Region*, USDA Occasional paper 27 (Southern Forest Experiment Station, US Forest Service,1932).

15. D. Sparks, "Adaptability of Pecan as a Species," *HortScience* 40 (2005): 1175–89.

16. Ibid.

17. M. F. Allen, *The Ecology of Mycorrhizae* (Cambridge: Cambridge University Press, 1991).

18. M. W. Smith, "Partitioning Phosphorus and Potassium in Pecan Trees during High and Low Crop Seasons," *Journal of the American Society for Horticultural Science* 134 (2009): 399–404.

19. R. E. Worley, "Effects of N, P, K, and Lime on Yield, Nut Quality, Tree Growth, and Leaf Analysis of Pecan," *Journal of the American Society for Horticultural Science* 99 (1974): 49–57.

20. M. L. Wells, "Response of Pecan Orchard Soil Chemical and Biological Quality Indicators to Poultry Litter Application and Clover Cover Crops," *HortScience* 46 (2011): 306–10.

21. Ibid.

22. D. Sparks, "Adaptability of Pecan as a Species," *HortScience* 40 (2005): 1175–89.

23. Ibid.

24. J. G. Woodruff and N. C. Woodruff, "Pecan Root Growth and Development," *Journal of Agricultural Research* 49 (1934): 511–30.

25. D. H. Marx and W. C. Bryan, "*Scleroderma bovista*, an Ectotrophic Mychorrizal Fungus of Pecan," *Phytopathology* 59 (1969): 1128–32.

26. G. Bonito, T. Brenneman, and R. Vilgalys, "Ectomycorrhizal Fungal Diversity in Orchards of Cultivated Pecan," *Mycorrhiza* 21 (2011): 601–12.

27. J. M. Trappe, "A. B. Frank and Mycorrhizae: The Challenge of Evolutionary and Ecological Theory," *Mycorrhiza* 15 (2005): 277–81.

28. C. Heimsch, "The First Recorded Truffle from Texas," *Mycologia* 50 (1958): 657–60.

29. G. Bonito, T. Brenneman, and R. Vilgalys, "Ectomycorrhizal Fungal Diversity in Orchards of Cultivated Pecan," *Mycorrhiza* 21 (2011): 601–12.

30. A. W. White and J. H. Edwards, "Soil and Root Growth Studies with Pecans at Byron, Georgia," *Pecan South* 13 (1980): 14–18.

31. Ibid.

32. D. Sparks, "Adaptability of Pecan as a Species," *HortScience* 40 (2005): 1175–89.

33. B. W. Wood, L. J. Grauke, and J. R. Payne, "Provenance Variation in Pecan," *Journal of the American Society for Horticultural Science* 123 (1998): 1023–28.

34. B. W. Wood, "Source of Pollen, Distance from Pollinizer, and Time of Pollination Affect Yields in Block-Type Pecan Orchards," *HortScience* 32 (1997): 1182–85.

35. B. W. Wood, "Pollen Characteristics of Pecan Trees and Orchards," *HortTechnology* 10 (2000): 120–126.

36. B. W. Wood, "Source of Pollen, Distance from Pollinizer, and Time of Pollination Affect Yields in Block-Type Pecan Orchards," *HortScience* 32 (1997): 1182–85.

37. D. Sparks and J. L. Heath, "Pistillate Flower and Fruit Drop of Pecan as a Function of Time and Shoot Length," *HortScience* 7 (1972): 402–4.

38. B. W. Wood and R. D. Marquard, "Estimates of Self-Pollination in Pecan Orchards in the Southeastern United States," *HortScience* 27 (1992): 406–8.

39. F. W. Went, "On Growth-Accelerating Substances in the Coleoptile of *Avena sativa*," *Proceedings of the Koninklijke Nederlandse Academie van Wetenschappen* 30 (1926): 10–19.

40. P. H. Raven, R. F. Evert, and S. E. Eichorn, *Biology of Plants* (New York: Worth Publishers, 1986).

41. L. Lombardini, H. Restrep-Diaz, and A. Volder, "Photosynthetic Light Response and Epidermal Characteristics of Sun and Shade Pecan Leaves," *Journal of the American Society for Horticultural Science* 134 (2009): 372–78.

42. D. Sparks, "Adaptability of Pecan as a Species," *HortScience* 40 (2005): 1175–89.

43. Ibid.

44. F. R. Brison, *Pecan Culture* (College Station: Texas Pecan Growers Association, 1974).

45. D. Sparks, "Adaptability of Pecan as a Species," *HortScience* 40 (2005): 1175–89.

46. M. J. Reigosa, N. Pedrol, and L. Gonzalez, *Allelopathy: A Physiological Process with Ecological Implications* (Dordrecht, the Netherlands: Springer, 2005).

47. P. R. Hoy and J. S. Stickney, "Comments on Walnut," *Transactions of the Wisconsin State Horticultural Society* 11 (1881): 166–67.

48. K. B. Coder, *Black Walnut Allelopathy: Tree Chemical Warfare*, University of Georgia, Warnell School of Forestry and Natural Resources Bulletin 11–10 (2011).

49. V. E. Langhans, P. A. Hedin, and C. H. Graves, "Fungitoxic Chemicals in Pecan Tissue," *Plant Disease Reporter* 2 (1978): 894–98.

50. M. W. Smith, "Understanding Alternate Bearing," *Pecan South* 38 (2005): 32–37.

51. M. W. Smith and J. C. Gallot, "Mechanical Thinning of Pecan Fruit," *HortScience* 25 (1990): 414–16.

52. D. Sparks, "A Climatic Model for Pecan Production under Humid Conditions," *Journal of the American Society for Horticultural Science* 121 (1996): 908–14.

53. W. D. Koenig and J. M. H. Knops, "The Mystery of Masting in Trees," *American Scientist* 93 (2005): 340–47.

CHAPTER 4

1. J. Graves, *Goodbye to a River* (New York: Knopf, 1960).

2. H. M. Dixon and H. W. Hawthorne, *An Economic Study of Farming in Sumter County, Georgia*, US Department of Agriculture Bulletin 492 (Washington, DC: US Department of Agriculture, 1917).

3. H. P. Stuckey and E. J. Kyle, *Pecan Growing* (New York: Macmillan, 1925).

4. Ibid.

5. W. Range, *A Century of Georgia Agriculture* (Athens, University of Georgia Press, 1954).

6. *History and Reminiscences of Dougherty County, Georgia* (Albany, GA: Thronateeska Chapter, Daughters of the American Revolution, 1924).

7. M. J. Cole, *From Stage Coaches to Train Whistles: History of Gum Pond, Mt. Enon, and Baconton in Mitchell County, Georgia* (Baconton, GA: Mt. Enon Historical Committee, 1977).

8. Ibid.

9. J. B. Wight, "Caring for the Pecan Orchard," *Proceedings, National Nut Growers Association* 5 (1906): 64–66.

10. J. B. Wight, *The Pecan: Some Points, Pointers, and Suggestions* (Jacksonville, FL: Record Company, 1916).

11. H. C. White, "Unscrupulous Dealers," *Nut Grower* 1 (1902): 1.

12. R. J. Redding, letter, *Nut Grower* 1 (1902): 6.

13. DuPont Company, *Farming with Dynamite: A Few Hints to Farmers* (Baltimore, MD: Lord Baltimore Press, 1911).

14. Ibid.

15. A. J. Farley, "Further Results with Dynamite for Tree Planting," *Proceedings, American Society for Horticultural Science* 11 (1914): 127–30.

16. W. A. Orton and F. V. Rand, "Pecan Rosette," *Journal of Agricultural Research* 3 (1914): 149–174.

17. F. R. Brison, "The History of Pecan Rosette," *Proceedings, Texas Pecan Growers Association* 44 (1965): 75–76.

18. A. O. Alben, J. R. Cole, and R. D. Lewis, "Chemical Treatment of Pecan Rosette," *Phytopathology* 22 (1932): 595–601.

19. A. O. Alben, J. R. Cole, and R. D. Lewis, "New Developments in Treating Pecan Rosette with Chemicals," *Phytopathology* 22 (1932): 979–81.

20. A. H. Finch, and A. F. Kinnison, *Pecan Rosette: Soil, Chemical, and Physiological Studies*, University of Arizona Agricultural Experiment Station Technical Bulletin 47 (1933).

21. D. Sparks, "Adaptability of Pecan as a Species," *HortScience* 40 (2005): 1175–89.

22. J. B. Storey, P. N. Westfall, and M. W. Smith, "Why Do Pecans Need Zinc?" *Pecan Quarterly* 13 (1979): 3–9.

23. P. F. Bertrand, "Pecan Scab," in *Compendium of Nut Crop Diseases in Temperate Zones*, edited by B. L. Teviotdale, T. J. Michailides, and J. W. Pscheidt (St. Paul, MN: American Phytopathological Society Press, 2002).

24. C. S. Spooner and C. G. Crittenden, "Pecan Diseases Other Than Scab," *Georgia State Board of Entomology Bulletin* 49 (1918): 38–48.

25. J. B. Demaree, "Latest Developments in the Control of Pecan Scab," *Proceedings, National Nut Growers Association* 21 (1922): 16–23.

26. P. M. A. Millardet, *The Discovery of Bordeaux Mixture*, trans. F. J. Schneiderhan (St. Paul, MN: American Phytopathological Society Press, 1885).

27. K. Robison, "100 Years of John Bean Spray Pump Company," *Gas Engine Magazine*, August/September 1985.

28. Ibid.

29. J.B. Demaree and J. R. Cole, "Dusting vs. Spraying for Pecan Scab," *Proceedings, National Pecan Growers Association* 25 (1926): 47–53.

30. R. D. Fox, R. C. Derksen, H. Zhu, R. D. Brazee, and S. A. Svensson, "A History of Air Blast Sprayer Development and Future Prospects," *Transactions of the American Society of Agricultural and Biological Engineers* 51 (2008): 405–10.

31. H. P. Stuckey and E. J. Kyle, *Pecan Growing* (New York: Macmillan, 1925).

32. H. E. Van Deman, "Comments on Pecan Inter-Cropping," *Proceedings, National Nut Growers Association* 5 (1906): 52.

33. J. B. Wight, "Caring for the Pecan Orchard," *Proceedings, National Nut Growers Association* 5 (1906): 64–66.

34. C. E. Pleas Plant Company, "All about Kudzu," *Nut Grower* 14 (1915): 28.

35. R. Y. Bailey, *Kudzu for Erosion Control in the Southeast*, USDA Farmers Bulletin no. 1840 (Washington, DC: US Department of Agriculture, 1944).

36. W. H. Harris, "Is the Inter-Cropping of Pecan Trees with Peach Trees Profitable?" *Proceedings, National Nut Growers Association* 21 (1922): 52–54.

37. R. E. Worley, "Meet Dr. Voigt: Grower, Experimenter, Breeder," *Pecan Quarterly* 5, no.1 (1971): 23.

38. G. R. McEachern, *Texas Pecan Handbook*, Texas Agricultural Extension Horticulture Handbook 105 (1997).

39. A. Winkler, "A Brief History of the Pecan Industry in Texas," *Proceedings, Texas Pecan Growers Association* 18 (1938): 64–69.

40. G. R. McEachern, *A Pecan History* (College Station: Texas A&M University, 2010).

41. J. R. Flack, "The Spread and Domestication of the Pecan in the U.S." (PhD diss., University of Wisconsin, 1970).

42. G. R. McEachern, *A Pecan History* (College Station: Texas A&M University, 2010).

43. Ibid.

44. Texas Department of Agriculture, *Pecans and Other Nuts in Texas*, Texas Department of Agriculture Bulletin No. 2 (1908).

45. W. J. Millican, "A Biographical Sketch of H. A. Halbert of Coleman, Texas," *Proceedings, Texas Pecan Growers Association* 18 (1938): 58–63.

46. "Pecan Budding in Seguin," *Seguin (TX) Enterprise*, July 5, 1907.

47. H. A. Halbert, "Originator of the Halbert Pecan Recounts Its Virtues," *American Nut Journal* 23 (1923): 87.

48. G. R. McEachern, *A Pecan History* (College Station: Texas A&M University, 2010).

49. H. P. Stuckey and E. J. Kyle, *Pecan Growing* (New York: Macmillan, 1925).

50. F. R. Brison, *Pecan Culture* (College Station: Texas Pecan Growers Association, 1974).

51. J. Manaster, *Pecans: The Story in a Nutshell* (Lubbock: Texas Tech University Press, 2008).

52. B. Waddell, "Burkett Origin Told," *Pecan Quarterly* 10 (1976): 8.

53. E. Oppenheimer, *Gilbert Onderdonk: The Nurseryman of Mission Valley* (Denton: University of North Texas Press, 1991).

54. G. Onderdonk, "Pomological Possibilities of Texas," *Texas Department of Agriculture Bulletin* 9 (1911): 18.

55. R. J. Bacon, "The Future of Pecans," *Proceedings, National Nut Growers Association* 20 (1921): 39–42.

56. S. Postel, *Pillar of Sand* (New York: Norton, 1999).

57. E. H. Davis, "Irrigation of Pecans," *Proceedings, Southeastern Pecan Growers Association* 40 (1947): 74–77.

58. A. O. Alben, "Results of an Irrigation Experiment on Stuart Pecan Trees in East Texas in 1956," *Proceedings, Texas Pecan Growers Association* 36 (1957): 16–23.

59. W. J. Florkowski, G. D. Humphries, and T. F. Crocker, *Commercial Pecan Tree Inventory, Georgia, 1997*, University of Georgia, College of Agricultural and Environmental Sciences, Agricultural Experiment Station Research Report 678 (2001).

60. E. D. DeRemer, "Does Trickle Irrigation Work in Pecan Orchards?" *Proceedings, Western Pecan Conference* 7 (1973): 91–92.

61. S. Blass, *Water in Strife and Action* (Givatayim, Israel: Masada, 1973).

62. E. D. DeRemer, "Does Trickle Irrigation Work in Pecan Orchards?" *Proceedings, Western Pecan Conference* 7 (1973): 91–92.

63. M. J. Keyworth, "Poverty, Solidarity, and Opportunity: The 1938 San Antonio Pecan Shellers' Strike" (master's thesis, Texas A&M University, 2007).

64. G. R. McEachern, *Texas Pecan Handbook*, Texas Agricultural Extension Horticulture Handbook 105 (1997).

65. Anonymous, "Louis D. Romberg: 1959–1960 President of Texas Pecan Growers Association," *Proceedings, Texas Pecan Growers Association* 38 (1959): 8–9.

66. L. D. Romberg, "Pecan Breeding and Related Problems," *Proceedings, Texas Pecan Growers Association* 44 (1965): 44–48.

67. R. D. Romberg, "Every Variety Has Own Special Problems," *Pecan Quarterly* 11 (1977): 18–19.

68. D. Sparks, *Pecan Cultivars* (Watkinsville, GA: Pecan Production Innovations, 1992).

69. L. J. Grauke, personal communication, May 22, 2012.

70. D. Sparks, *Pecan Cultivars* (Watkinsville, GA: Pecan Production Innovations, 1992).

71. E. Herrera, *Historical Background of Pecan Plantings in the Western Region*, Guide H-626 PHI–110, (New Mexico State University, 2000).

72. D. A. Fryxell, "The Red or Greening of New Mexico," *Desert Exposure*, December 2007, http://www.desertexposure.com/200712/200712_garcia_chile.php.

73. N. L. Newcomer, "Tribute to a Leader . . . Deane Stahmann," *Pecan Quarterly* 4, no. 4 (1970): 24–25.

74. Ibid.

75. G. Duffy, obituary, R. Keith Walden, *Tucson (AZ) Citizen*, March 15, 2002.

76. Testimony of Nan Stockholm Walden before the Subcommittee on National Parks, Forests, and Public Lands, US House Committee on Natural Resources on HR 2944, Southern Arizona Public Lands Protection Act, January 21, 2010.

77. K. Nuzam, *A History of Baldwin County* (Fairhope, AL: Page & Palette, 1971).

78. C. H. Graves, C. Hines, and W. W. Neel, "Camille Deals Heavy Blow to Mississippi Groves," *Pecan Quarterly* 4 (1970): 5–6.

79. T. B. Hagler, J. L. Boutwell, and L. Johnson, "Damage to Pecans in Alabama by Hurricane Frederic," *Proceedings of Southeastern Pecan Growers Association* 73 (1980): 99–105.

80. B. W. Wood, W. Goff, and M. Nesbitt, "Pecans and Hurricanes," *HortScience* 36 (2001): 253–58.

81. "Our History," The Samuel Roberts Noble Foundation, Ardmore, OK, accessed July 27, 2012, http://www.noble.org/about/history/.

82. W. Reid, and K. L. Hunt, "Pecan Production in the Northern United States," *HortTechnology* 10 (2000): 298–301.

83. Ibid.

84. F. Boyett, "Shotgun Wedding Sprouted Famous Pecan," *Evansville (IN) Courier Press*, February 11, 2012.

85. W. Reid and K. L. Hunt, "Pecan Production in the Northern United States," *HortTechnology* 10 (2000): 298–301.

Chapter 5

1. E. Oppenheimer, *Gilbert Onderdonk: The Nurseryman of Mission Valley* (Denton: University of North Texas Press, 1991).

2. F. R. Brison, *Pecan Culture* (College Station: Texas Pecan Growers Association, 1974).

3. G. R. McEachern, "Castro Provided Unequaled Contribution to Mexican Industry," *Pecan South* 43, no. 4 (2010): 29–30.

4. L. Lombardini, "Nogatec Provides Forum for Mexico's Producers," *Pecan South* 37, no. 7 (2004): 24–25.

5. F. R. Brison, *Pecan Culture* (College Station: Texas Pecan Growers Association, 1974).

6. L. Lombardini, "Nogatec Provides Forum for Mexico's Producers," *Pecan South* 37, no. 7 (2004): 24–25.

7. F. R. Brison, *Pecan Culture* (College Station: Texas Pecan Growers Association, 1974).

8. C. L. Wise, "Jimenez Conference Attracts Large Crowd," *Pecan South* 44, no. 9 (2011): 20–22.

9. S. Vazquez, "The Walnut Trees of El Valle," *El Sol de Parral*, Parral, Chihuahua, Mexico. July 28, 2009.

10. E. Herrera, and L. C. Velo-Duran, "Mexico's Stately Natives Beleaguered by Nature," *Pecan Quarterly* 18, no. 3(1984): 5–6.

11. USDA-FAS, *Mexico Tree Nuts Annual*, GAIN Report no. MX9061 (2009).

12. E. C. Harter, *The Lost Colony of the Confederacy* (College Station: Texas A&M University Press, 2000).

13. Ibid.

14. G. Hawkins, "The Lost Confederados," *Garden and Gun*, September/October 2008.

15. R. Harris, "In Brazil, Heirs Recall Last Stand of U.S. Confederacy," *Los Angeles Times*, April 17, 1995.

16. E. C. Harter, *The Lost Colony of the Confederacy* (College Station: Texas A&M University Press, 2000).

17. G. Hawkins, "The Lost Confederados," *Garden and Gun*, September/October 2008.

18. D. L. Stepp and R. Vise, "Stahmann Farms Produce Pecans on Two Hemispheres," EPCC *Borderlands*, accessed March 12, 2012, http://epcc.libguides.com/content.php?pid=309255&sid=2891582.

19. E. A. Frusso, personal communication, May 30, 2012.

20. Ibid.

21. F. R. Brison, *Pecan Culture* (College Station: Texas Pecan Growers Association, 1974).

22. Ibid.

23. D. L. Stepp and R. Vise, "Stahmann Farms Produce Pecans on Two Hemispheres," EPCC *Borderlands*, accessed March 12, 2012, http://epcc.libguides.com/content.php?pid=309255&sid=2891582.

24. B. W. Wood and D. F. Stahmann, "Hedge Pruning Pecan," *HortTechnology* 14 (2004): 63–72.

25. D. L. Stepp and R. Vise, "Stahmann Farms Produce Pecans on Two Hemispheres," EPCC *Borderlands*, accessed March 12, 2012, http://epcc.libguides.com/content.php?pid=309255&sid=2891582.

26. F. R. Brison, *Pecan Culture* (College Station: Texas Pecan Growers Association, 1974).

27. B. Guest, *A Fine Band of Farmers: A History of Agricultural Studies in Pietermaritzburg* (Pietermaritzburg, South Africa: Natal Society Foundation, 2010).

28. F. R. Brison, *Pecan Culture* (College Station: Texas Pecan Growers Association, 1974).

29. Ibid.

30. J. Seshoka, Willem De Lange, Nicolas Faysse, *The Transformation of Irrigation Boards into Water User Associations in South Africa* (International Water Use Institute, 2004).

31. S. Homsky, "The Pecan Growing Industry in Israel," *Pecan South* 4, no. 4 (1977): 178–85.

32. U. Yermiyahu, A. Tal, A. Ben-Gal, A. Bar-Tal, J. Tarchitzky, and O. Lahav, "Rethinking Desalinated Water Quality and Agriculture," *Science* 318 (2007): 920–21.

33. J. Fedler, "Israel's Agriculture in the 21st Century," Israel Ministry of Foreign Affairs, 2002, accessed June 10, 2012, http://www.mfa.gov.il/MFA/Facts+About+Israel/Economy/Focus+on+Israel-+Israel-s+Agriculture+in+the+21st.htm.

34. U. Yermiyahu, A. Tal, A. Ben-Gal, A. Bar-Tal, J. Tarchitzky, and O. Lahav, "Rethinking Desalinated Water Quality and Agriculture," *Science* 318 (2007): 920–21.

35. Ibid.

36. "Apollo 16 Mission Commentary," National Aeronautics and Space Administration, Washington, DC, 1972, accessed August 28, 2012, http://www.jsc.nasa.gov/history/mission_trans/AS16_PAC.PDF.

37. "Apollo 17 Press Kit," National Aeronautics and Space Administration, Washington, DC, 1972, accessed August 28, 2012, http://history.nasa.gov/alsj/a17/A17_PressKit.pdf.

CHAPTER 6

1. M. A. Altieri, "The Ecological Role of Biodiversity in Agroecosystems," *Agriculture, Ecosystems, and Environment* 74 (1999): 19–31.

2. C. R. Clement, W. M. Denevan, M. J. Heckenberger, A. B. Junqueira, E. G. Neves, W. G. Teixeira, and W. I. Woods, "The Domestication of Amazonia before European Conquest," *Proceedings of The Royal Society B* 282 (2015): 813.

3. S. Simion, J. C. Bouvier, J. F. Debras, and B. Sauphanor, "Biodiversity and Pest Management in Orchard Systems: A Review," *Agronomy for Sustainable Development* 30 (2010): 139–52.

4. R. E. Worley, *Compendium of Pecan Production and Research* (Ann Arbor, MI: Edwards Brothers Press, 2002).

5. M. W. Smith and B. W. Wood, "Pecan Tree Biomass Estimates," *HortScience* 41 (2006): 1286–91.

6. M. L. Wells, "Pecan Nutrient Element Status and Orchard Soil Fertility in the Southeastern United States Coastal Plain," *HortTechnology* 19 (2009): 432–38.

7. M. W. Smith and B. W. Wood, "Pecan Tree Biomass Estimates," *HortScience* 41 (2006): 1286–91.

8. V. Smil, *Enriching the Earth* (Cambridge, MA: MIT Press, 2001).

9. Sustainable Agriculture Network, *Managing Cover Crops Profitably*, Handbook Series (1998).

10. J. D. Dutcher, M. L. Wells, T. B. Brenneman, and M. G. Patterson, "Integration of Insect and Mite, Disease, and Weed Management to Improve Pecan Production," in *Integrated Pest and Disease Management*, vol. 5, edited by A. Ciancio and K. G. Mukerji (Dordrecht, the Netherlands: Springer, 2009).

11. M. L. Wells, "Response of Pecan Orchard Soil Chemical and Biological Quality Indicators to Poultry Litter Application and Clover Cover Crops," *HortScience* 46 (2011): 306–10.

12. M. L. Wells, "Pecan Response to Nitrogen Fertilizer Placement," *HortScience* 48 (2013): 369–72.

13. L. E. Acuña-Maldonado, M. W. Smith, N. O. Maness, B. S. Cheary, and B. L. Carroll, "Influence of Nitrogen Application Time on Nitrogen Absorption, Partitioning, and Yield of Pecan," *Journal of the American Society for Horticultural Science* 128 (2003): 155–62.

14. M. L. Wells, "Pecan Response to Nitrogen Fertilizer Placement," *HortScience* 48 (2013): 369–72.

15. S. C. Allen, S. Jose, P. K. R. Nair, B. J. Brecke, P. Nkedi-Kizza, and C. L. Ramsey, "Safety Net Role of Tree Roots: Evidence from a Pecan-Cotton Alley Cropping System in the Southern United States," *Forest Ecology and Management* 192 (2004): 395–407.

16. J. D. Dutcher, M. L. Wells, T. B. Brenneman, and M. G. Patterson, "Integration of Insect and Mite, Disease, and Weed Management to Improve Pecan Production," in *Integrated Pest and Disease Management*, vol. 5, edited by A. Ciancio and K. G. Mukerji (Dordrecht, the Netherlands: Springer, 2009).

17. Ibid.

18. M. A. Altieri, "The Ecological Role of Biodiversity in Agroecosystems," *Agriculture, Ecosystems, and Environment* 74 (1999): 19–31.

19. S. Simion, J. C. Bouvier, J. F. Debras, and B. Sauphanor, "Biodiversity and Pest Management in Orchard Systems: A Review," *Agronomy for Sustainable Development* 30 (2010): 139–52.

20. J. D. Dutcher, M. L. Wells, T. B. Brenneman, and M. G. Patterson, "Integration of Insect and Mite, Disease, and Weed Management to Improve Pecan Production," in *Integrated Pest and Disease Management*, vol. 5, edited by A. Ciancio and K. G. Mukerji (Dordrecht, the Netherlands: Springer, 2009).

21. Ibid.

22. A. Calixto, A. Dean, A. Knutson, M. K. Harris, and B. Ree, "Spiders in Texas Pecans, Are They Affected by Fire Ants?" *Newsletter of the American Arachnological Society* 66 (2003): 4.

23. J. D. Dutcher, M. L. Wells, T. B. Brenneman, and M. G. Patterson, "Integration of Insect and Mite, Disease, and Weed Management to Improve Pecan Production," in *Integrated Pest and Disease Management*, vol. 5, edited by A. Ciancio and K. G. Mukerji (Dordrecht, the Netherlands: Springer, 2009).

24. R. F. Mizell, and D. E. Shiffhauer, "Seasonal Abundance of the Crapemyrtle Aphid in Relation to the Pecan Aphids and Their Common Predators," *Entomophaga* 32 (1987): 511–20.

25. S. Simion, J. C. Bouvier, J. F. Debras, and B. Sauphanor, "Biodiversity and Pest Management in Orchard Systems: A Review," *Agronomy for Sustainable Development* 30 (2010): 139–52.

26. Ibid.

27. W. T. Stamps and M. J. Linit, "Plant Diversity in Arthropod Communities: Implications for Temperate Agroforestry," *Agroforestry Systems* 39 (1998): 73–89.

28. D. H. Janzen, "Host Plants as Islands in Evolutionary and Contemporary Time," *American Naturalist* 102 (1968): 592–95.

29. D. H. Janzen, "Host Plants as Islands. II. Competition in Evolutionary and Contemporary Time," *American Naturalist* 107 (1973): 786–90.

30. G. Beaton, *Birding Georgia* (Helena, MT: Falcon Guides, 2000).

31. K. Blankenship, "Bulloch County Birding Locations," 2009, accessed July 23, 2012, http://www.wingsoverga.com/BullochCoBirdingSites.html.

32. J. L. Landers and R. L. Crawford, *Nongame Wildlife Habitat Management Demonstration Area*, Florida Game and Freshwater Fish Commission Nongame Wildlife Program Project (1995).

33. K. J. Kingsley, 1985. "The Pecan Orchard as a Riparian Ecosystem" (paper presented to the Interagency North American Riparian Conference on Riparian Ecosystems and Their Management, 1985), http://www.fs.fed.us/rm/pubs_rm/rm_gtr120/rm_gtr120_245_249.pdf.

34. R. Carson, *Silent Spring* (New York: Houghton Mifflin, 1962).

CHAPTER 7

1. M. L. Wells, "Cost of Pecan Production," *Pecan Grower* 20, no. 2 (2008): 30–31.

2. D. Wessel, "Shell Shock: Chinese Demand Reshapes U.S. Pecan Business," *Wall Street Journal*, April 18, 2011.

3. C. Mann, *1493: Uncovering the New World Columbus Created* (New York: Vintage, 2011).

4. D. Wessel, "Shell Shock: Chinese Demand Reshapes U.S. Pecan Business," *Wall Street Journal*, April 18, 2011.

5. Z. J. Sun and S. A. He, "The History, Present, and Prospect of Pecan in China," *Pecan South* 9, no. 5 (1982): 18–23.

6. Ibid.

7. L. J. Grauke, "Yunnan Pecans," 2005, http://aggie-horticulture.tamu.edu/carya/cgc/Yunnan%20Pecans.pdf.

8. Ibid.

9. "Mediterranean Diet: A Heart-Healthy Eating Plan," Mayo Clinic, http://www.mayoclinic.org/healthy-lifestyle/nutrition-and-healthy-eating/in-depth/mediterranean-diet/art-20047801.

10. L. R. Beuchat and R. E. Worthington, "Technical Note: Fatty Acid Composition of Tree Nut Oils," *Journal of Food Technology* 13 (1978): 355–58.

11. E. Ros, "Health Benefits of Nut Consumption," *Nutrients* 2 (2010): 652–82.

12. R. R. Eitenmiller and R. B. Pegg, "Compositional Characteristics and Health Effects of Pecan [*Carya illinoinensis* (Wangenh.) K. Koch]," in *Tree Nuts: Composition, Phytochemicals, and Health Effects*, edited by C. Alasalvar and F. Shahidi (Boca Raton, FL: CRC Press, 2008).

13. S. Rajaram, K. Burke, B. Connell, T. Myint, and J. Sabaté, "A Monounsaturated Fatty Acid-Rich Pecan-Enriched Diet Favorably Alters the Serum Lipid Profile of Healthy Men and Women," *Journal of Nutrition* 131 (2001): 2275–79.

14. X. Wu, G. R. Beecher, J. M. Holden, D. B. Haytowitz, S. E. Gebhardt, and R. L. Prior, "Lipophilic and Hydrophilic Antioxidant Capacities of Common Foods in the United States," *Journal of Agricultural and Food Chemistry* 52 (2004): 4026–37.

15. R. R. Eitenmiller and R. B. Pegg, "Compositional Characteristics and Health Effects of Pecan [*Carya illinoinensis* (Wangenh.) K. Koch]," in *Tree Nuts: Composition, Phytochemicals, and Health Effects*, edited by C. Alasalvar and F. Shahidi (Boca Raton, FL: CRC Press, 2008).

16. Ibid.

17. Ibid.

18. Y. Bao, J. Han, F. B. Hu, E. L. Giovannucci, M. J. Stampfer, W. C. Willett, and C. S. Fuchs, "Association of Nut Consumption with Total and Cause Specific Mortality," *New England Journal of Medicine* 369 (2013): 2001–11.

19. M. L. Wells, "Pecan Planting Trends in Georgia," *HortTechnology* 24 (2014): 475–79.

20. Ibid.

21. Ibid.

Index

acorns, 13

African Americans, xxiii

Alabama, 9, 10, 47–48, 162–64, 165–66

Albany District Pecan Exchange, 103–4

Algonquin Indians, 5

allelopathy, 89

almonds, xxii, xxiii, 100, 119, 223

Amling, Harry, 147, 181

Antoine, 38, 39, 40, 41, 42, 60, 145

arborvitae, xxiv

Argentina, 182–84

Arizona, 61, 148, 159–60

Arkansas, 22, 102

Australia, 49, 184–87

Bacon, George Meriweather, 104, 108, 109, 110

Bacon, R. J., 104, 108

Bailey, Liberty Hyde, 34

Bartram, John, 23–25

Bay View Nursery, 56

Bechtel, Theodore, 58–60

Bertrand, Paul, 125

Blackmon, G. H., 131

Blass, Simcha, 137–38, 191

Bonzano, Huberto, 39, 40

botanists, leading English and American, 6

Bourgeois, Emil, 42, 43, 60

Brazil, 178–82, 184

Brenneman, Tim, 79

Brison, Fred, 65, 185

budding, 33–34, 35, 36, 37–38, 110

Burkett, James Henry, 25, 131–33, 146

butternuts, 13

Cabeza de Vaca, Álvar Núñez, xiv, xxiv, 2, 14, 18, 21, 22

Caddo Indians, 17

California, 148

Canada, 217

casket girls, xxiii

Castanera, Eugene, 48, 49

Castro Medina, Ruben, 174, 188

Centennial Exposition, 39–40

Charlevoix, Xavier, 4–5, 22

Cherokee Indians, 20

China, 1, 8, 35, 184, 186, 187, 213–17, 221, 223

Choctaw Indians, 10

climate, 61, 81, 83, 94–95, 148, 169, 181, 204; cold, 10, 48, 62–63, 64, 97, 170; dry (see water); hot, 63–64; 66–68, 80; wet (see water)

Colomb, A. E., 38–39, 40

Comanche Indians, 16

congealed fruit salad, 231

Connecticut, 64

Conner, Patrick, 115, 151, 206

Corn Products Company, xxi

Coronado, Francisco, 21–22

cotton, 103, 121, 127–28

Creek Indians, 20, 162

Creole culture, xxiii

Crocker, Tom, 125

Crockett, David, 20–21

crop size, 48, 58, 59, 65–66, 91, 94, 144, 177, 186, 197

Dalton Culture, 13

Delmas, Albert Grant, 48, 49, 53, 54

Dorsey, George, 17

Doyle, A. T., 185

drought. *See* water

Duerler, Gustav Anton, 140–41

Ellis, H. C., 125

Europe, 217

Evans, J. A., 131

Fairchild, David, 26

Farley, A. J., 111

Farmers Investment Company (FICO), 158, 159, 160–61

fertilizer, xviii, xix, 73–74, 100, 103, 105, 112, 143, 166, 201, 203; commercial, 75, 196, 198; natural, 156; timing of, 202; water-soluble, 138, 139

filberts, 100

Flagg, Edmund, 23

Florida, xiii–xiv, 9, 54, 78

Food Machinery Corporation, 119

Forkert, Charles Augustus, 55–57, 58

Frank, Albert Bernhard, 78

French settlers, xxii–xxiii, 3–5, 23, 30, 46

fungi, 72, 77–79, 89, 112, 144, 201. *See also* pecan scab

Garcia, Fabian, 152–55, 156

Georgia, 9, 49, 78; as the leading US pecan producer, xxv, 102–8, 114, 122–25, 168, 221–22; pecan consumption in, 101; pecan orchards in, 75, 77, 79, 87, 137, 150, 160, 199, 204, 207, 211; pecan varieties in, 54–55, 57, 59, 116, 140, 148, 149, 150, 151; predicting pecan production in, 95

Georgia Agricultural Experiment Station, 110, 121, 131

grafting, 33–34, 35, 36, 38, 42, 43, 45, 48, 62, 110, 123, 146

Grape-Nuts cereal, 45

Grauke, L. J., 11–12, 149–50

Graves, John, 101

Greber, Norman, 185

green gelatin salad, 235

Halbert, Halkert A., 128–30, 131

Harris, W. H., 122

Harrison, Vesta, xx–xxi

hazelnuts, 13

hedging, xviii–xix, 148, 157–58, 160

Herring, Mrs. Frank, xxi

hickory nuts, xx, 2–3, 19; origins of, 8, 11, 12–14

hurricanes, 164–66

Iguace Indians, 18

Illinois, 12, 14, 150, 169; Kaskaskia, 4, 22–23, 30

India, 217

Indiana, 9, 169, 170

insecticides. *See* pecan orchards, pest management in

Iowa, 9, 10, 13, 61, 62, 169

irregular bearing, 91–92, 94, 167

irrigation, 51, 68, 94, 105, 133, 143; amount of water allocated for, 161–62; canals and ditches, 25, 134, 154, 175; deficit, 174; drip, xix, 137–38, 163, 175, 186, 191; engineering, 187; flood, 134–35, 136, 175, 186; furrow, 135; geography and need for, 99; importance of, 134, 139, 152; lines, 197; maximizing efficiency of, 161; overhead, 115; scheduling, 100; sprinkler, xix, 136–37, 138; and water quality, 156, 189

Israel, 137, 190–92

Jackson Pecan Grove Company, 104

Jefferson, Thomas, 28–30

Jumano Apache Indians, 18, 20

Kansas, 65, 144, 169

Karo corn syrup, xxi, xxii

Keller, John, 49

Kentucky, 10

Keystone Pecan Company, xxv

Kiowa Indians, 16

Koch, Karl, 7

kudzu, 121–22

Kyle, E. J., 130

Landrum, Abner, 36–38

Lassabe, John R., 47, 49

Lassagne, Clement, xxii

lemon broccoli with pecans, 227

León, Ponce de, 23

Logan, James, 20

Louisiana, 12, 19, 22, 144, 164–66; New Orleans, xxii–xxiii, 31–32, 47, 208 (*see also* Oak Alley Plantation); Saint James Parish, 39, 42, 43

macadamia nuts, 100

Madden, George, 147

Maison de la Praline, xxii

Marest, Gabriel, 4, 22

Mariame Indians, xiv, 18

Marshall, Humphrey, 6

masting, 91, 94, 95, 96

McCulloch, Ben, 20–21

Mendoza, Pedro de, 23

Mescalero Apache Indians, 18–19

Mexico, xvii, 1, 9–10, 20, 25, 61, 63, 64, 133, 150; legendary pecan trees of, 176–78; and pecan DNA, 11–12; as the second leading world pecan producer, 173–78

Michaux, André, 30, 71

Middle East, 217; *see also* Israel

Minifie, George, 1

Mire, Duminie, 42–43

Mississippi, 3–4, 10, 42, 48, 54, 60, 144, 164–66; Coahoma County, 101–2; Jackson County, 46–47, 51, 55, 60; Ocean Springs, 43, 46, 47, 49, 50–51, 55–56, 58–59

Mississippi River Valley, 9, 10, 12–13, 16, 61, 68, 101

Missouri, 65, 144, 169

Moore, James, 50
Moran effect, 97
Murch, Sue, xx

Narváez, Pánfilo de: expedition of,
    xiii–xiv
NASA, 192–93
Natchez Indians, 3–4
National Clonal Germplasm Repository
    for Pecans and Hickories, 150
National Nut Growers Association, 56,
    109, 110, 121, 122, 134
National Tree Registry, 65
Native Americans, xiii, xiv, 10, 14, 19, 30,
    91, 224; discovery of pecan by, xvii–
    xviii; and name of pecan, 3; planting
    of pecans by, 13, 25; and reverence for
    pecan tree, 17; use of pecans for food
    and drink by, xx, xxi, 2, 16, 18; use of
    pecans for trade by, 19–20; varieties
    named for, 147–48. See also tribal names
New Jersey Experiment Station, 111
New Mexico, 61, 78, 148, 152–55, 161, 185; as
    the third leading US pecan producer,
    158, 168
New York, 31
nitrogen, 74–75, 77, 80, 198, 199–200, 201,
    202–3, 204
Niza, Marcos de, 21
North Carolina, 9, 61, 123, 204
nutmeg hickory, 10
Nuttall, Thomas, 2

Oak Alley Plantation, 32–33, 38–39, 40,
    41, 46

Ocean Springs Pecan Nursery, 50
Ohio, 9, 10, 71
Oklahoma, 9, 19, 101, 128, 144, 148, 166–68
Onderdonk, Gilbert, 133, 135, 173
Owens, Frank, 101

Pabst, Charles E., 50, 52, 59
Payaya Indians, 19
Pecan City, 107, 160
pecan-coated pork tenderloin, 232
pecan-encrusted fish, 229
Pecan Growers Associations, 123, 215, 217
pecan industry, xiv, xxv, 2, 37, 38, 42, 45,
    49, 57, 59, 109, 113, 122, 213; in Alabama,
    163–64, 166; in Australia, 49, 185–87; in
    the Desert Southwest, 68, 99–100,134,
    152, 155, 158, 203; frauds in, 110; future
    of, 223; in Georgia, 102–8, 124–25,
    221–22; in Israel, 192; machines in, 120,
    168; Mexican workers in, 140–43; in
    Mexico, 174, 175–76; in Mississippi, 47;
    in the Northern Pecan Region, 168–71;
    in Oklahoma, 166, 168; "package"
    approach in, 125; production costs of,
    213–14; rise of, 42, 46–47; in South
    Africa, 187–88; in Texas, 127–28, 130,
    131, 137, 139; threats to, 33, 115
Pecanita, 181, 182
pecan orchards, xviii, 40–41, 52, 73, 97;
    and agroecology, 195–212; in Alabama,
    163; in Argentina, 183; in Australia, 185;
    benefits of legumes (especially clover)
    in, xviii, 74–75, 121, 197, 199–201, 207;
    birds in, 87, 211; in Brazil, 181; crop
    interplanting in, 120–22, 127, 157,

163, 174, 197, 221; diversity in, 205–8, 209–11; in Georgia (*see* Georgia, pecan orchards in); or "grove," 110; and land schemes, 105–8; largest contiguous block of, 107; management of, 100, 105, 143–44; in Mexico, 173; in Oklahoma, 166; pest management in, xix, 55, 119–20, 100, 143, 181, 186, 203–5, 206–11; pollinators, 84, 96; requirements for, 101; sale of, 105; seedling, 143, 144; size of, 101, 186; in Texas, 8, 126, 150, 198; use of dynamite in, 110–11. *See also* pecan trees

pecan pie, xviii, 218; original version of, xx–xxi; recipes for, xx, xxi, 233; and Southern culture, xx, xxii, xxiii–xxiv (*see also state names*)

pecan recipes, xx, xxi, 225–35

pecan rosette, 112–14

pecans, xiv, xx, 2; adaptability of, 68; animal use of, 87–89; chopped, xxii; consumption of, 101; DNA of, 11–12; domestication of, 1–2; early cultivation of, 25–30; first known description of, xiv; human effect on, 7–21; international recognition of, 2; introduction into Europe of, 23; introduction on East Coast of, 24; name of, 2–7, 110; nutritional and health value of, xxiv, 18, 192–93, 217–21; role in diet of, xxiv; shelling (cracking) of, 140–43; toasted, 225; varieties of, 33, 46–47, 60, 115, 140, 143–54, 169 (*see also* pecan varieties, names of)

pecan scab, 53, 55, 57, 58, 59, 90, 115, 148, 149, 188, 206–7; and climate, xix, 165, 181, 186, 204; fungicides, xviii, 60, 116, 118–20, 205; and on Gulf Coast, 163

pecan trees, xxii, xxv; breeding of, 44, 55, 56–57, 59, 60, 115, 146–52, 206; defenses of, 89, 152; description of, xvii; flowers of, 81–83, 84, 92–93, 95, 96–97, 151; geographical range of, 9–10, 97, 99, 101, 185 (*see also state and country names*); growth cycle of, xviii–xix, 19, 62, 80, 84, 85, 90–92, 93, 95, 100, 105, 120; leaf shedding by, 68–69, 198, 199; life span of, 65; Mexican legendary, 176–78; nutritional requirements of, 80–81; rise in number of, 102; roots of, 68, 69, 70, 74, 75–77, 79–80, 93, 95, 198; seedlings of, 33, 36, 42, 43, 47, 62, 83–84, 86, 101, 123, 140, 143, 144, 146, 150, 152, 181, 216; size of, 65–66, 75, 148, 177; sporadic nature of, 33; as "tree of life," xxiv, 212, 224. *See also* pecan orchards

pecan truffle, 78–79

pecan varieties, names of: Admirable, 57; Alley, 47, 52–53, 146, 206; Amling, 164; Apache, 147; Barton, 146, 181, 182; Big Z, 190; Bradley, 158, 176; Brooks, 147; Burkett, 132–33, 147, 173, 175, 190; Byrd, 151; Caddo, 164, 166, 222; Candy, 60; Cape Fear, 163, 164, 181, 182, 184, 221; Castanera, 48; Centennial, 40–42; Cherokee, 148; Chetopa, 169; Cheyenne, 148, 163; Chickasaw, 148, 181, 182; Choctaw, 127, 188, 190; Clark, 147, 175; Colby, 169, 184; Columbia,

163; Comanche, 147; Cunard, 151; Curtis, 147; Delmas, 54, 115, 123, 190; Dependable, 56; Desirable, 56, 57–58, 59, 63, 88, 90, 107, 116, 146, 147, 163, 164, 181, 182, 184, 190, 192, 221; Elliot, 115, 163, 164, 175; Evers, 147; Excel, 164, 222; Faith, 169; Forkert, 164, 184; Frotscher, 40, 90, 109, 163, 173; Gafford, 164; Garner, 190; Giles, 169, 184; Gloria Grande, 184; Goosepond, 169; Govett, 190; Halbert, 129, 148, 190; Harris Super, 184; Hastings, 175; Hirschi, 169; Hodge, 184; Ideal, 175; Jackson, 60; Jayhawk, 169; Jewett, 56; Kanza, 166, 170; Kernoodle, 184; Kiowa, 166, 184; Lakota, 170; Lane, 188; Lerouk, 188; Lucas, 169, 184; Mahan, 63, 147, 148, 174, 175, 184, 185, 190, 215; Major, 147, 169–70, 184; McMillan, 164; Mobile, 163; Mohawk, 147, 148, 149; Moneymaker, 163, 173, 190; Moore, 147, 175; Morrill, 151; Navajo, 190; Nelis, 190; No. 60, 45, 175; Norton, 169; Oconee, 184, 222; Odum, 147; Onliwon, 45; Osage, 169, 184; Owens, 102; Pabst, 47, 52, 115, 123, 146, 185, 206; Pawnee, 59, 148, 149, 166, 169, 170, 184, 190, 221; Peruque, 169, 184; Rome, 40; Russell, 50, 57, 127; San Saba, 44–45, 146, 190; Schley, 47, 53–55, 63, 107, 115, 123, 146, 147, 164, 190, 206; Shawnee, 148, 181, 182; Shoshoni, 138, 181, 182, 184; Squirrel's Delight, 45, 175; Starking Hardy Giant, 147, 149, 184; Stevens, 90; Stuart, 47, 48–49, 53, 55, 58, 63, 107, 123,

146, 163, 164, 169, 173, 184, 185, 206, 222; Success, 56, 57, 59, 63, 146, 147, 163, 164, 173, 184; Sumner, 184, 221; Surprize, 164; Texas Prolific, 45; Ukulings, 188; Van Deman, 43, 115, 146, 173; Vlok, 188; Western Schley, 45, 158, 174, 175, 176, 181, 184, 185, 186, 190; Wichita, 148, 174, 175, 181, 184, 186, 188, 189, 190; Williamson, 185
pecan waffles, 226
pecan weevil, 170, 181
Penicaut, Andre, 4
Pennsylvania, 31
Perry, James, 20
Peru, 184
phosphorus, 73, 77, 80, 203
photosynthesis, 67, 86, 94, 95, 202
Pierson, D. L., 53–54
pistachios, 100, 223
pollen data, 11, 13
pollination, 45, 56, 63, 82–84, 96, 151, 206
powcohiccora (drink), xx, 16
pralines, xx, xxii–xxiii, 218; recipe for, 234
Prince Nursery, 26

Quevene Indians, xiv

rainfall. See water
Rawlings, Marjorie Kinnan, xxi
Redding, R. J., 110, 111
Risien, E. E., 42, 43–46, 56, 60, 127, 145
Road of Remembrance, 101
Romberg, Louis D., 146–48, 149, 150
Rose, Edgar, xxii
Russia, 217

Sabeata, Juan, 18

Schmidt, William B., 51–52, 59

Schmidt & Ziegler, 51

Scott, R. Lloyd, 123

Seligmann, Julius, 141–42

Seton, Ernest Thompson, 97

shagbark hickory, 10

shellbark hickory, 13

Smith, J. Russell, xxiv

Smith, Mike, 94, 166, 168

Snedeker, Albert Clark, 123

soil, 122, 135; acidity, 10, 81, 113–14; moisture, 10, 66–71, 75–77, 80, 94–95, 97, 100, 136, 144; organic matter in, 71–74, 80, 197–202; salinity, 139, 175, 190; and sustainability, 196–97; temperature, 79

Soto, Hernando de, 21, 22, 23

South Africa, 187–90

South Carolina, 9, 36, 123, 204

Southern Pecan Shelling Company, 141–42

South Korea, 217

Spanish explorers, xiii, 20, 21–23, 26

Sparks, Darrell, 87, 95, 151

Stahmann Farms, 155–58, 160, 185–87

stomata, 66, 67, 68

Storey, Benton, 113, 130

Stuart Pecan Company, 47

Stuart, William Rasin, 47–50, 184

Summit Nurseries, 54

Sunday chicken salad, 230

sunlight, 64, 69, 85, 86–87, 100–101, 114, 144, 198

sweet potato casserole, 228

Swinden, F. A., 127

Tenayuca, Emma, 142

Tennessee, 9, 65

Texas, 10, 12, 19, 21, 23, 36, 42, 49, 68, 101, 102, 112, 150, 204; Austin, 146; central, 9, 15, 20, 61, 62, 137, 152, 204; drought in, 76, 161; Edwards Plateau, 9, 14–16; pecan orchards in, 78, 126, 150, 198; Pecan Pie, xx–xxi; San Antonio, 139–43; San Saba, 43–44, 126; as the second leading US pecan producer, 125–33, 143, 144, 168; soil in, 81; southern, xvii, 1, 14; Weatherford, 65; West, 61, 133, 137, 148, 161

Texas A&M University, 65, 114, 121, 126, 130, 131, 146, 168, 188, 192

Texas Department of Agriculture, 25, 127, 131,132, 146

Texas Nut Growers Association, 126, 128

Teyas Indians, 22

Thompson, Tommy, 83, 149, 150

Tlaxcalan Indians, 25

top-working, 53, 57, 128, 129, 131, 149, 173

trees: benefits of, xxiv, 198; crop size of, 95–96; fruit, 26–27, 35, 62, 93; grafting of, 33–38; types that grow with pecan trees, 71, 185

tree shakers, 94, 168, 182

True, Rodney, 2

Turkey, 217

Twain, Mark, 31, 32

Uruguay, 184

US Department of Agriculture (USDA), 24, 43, 48, 57, 59, 83, 112, 133, 208; breeding programs, 55, 60, 146–49,

150–51, 163, 170, 188, 190; pecan field
stations, 11, 57, 146, 150, 215, 216

Virginia, 1, 31
Voigt, Fred, 123–24

Walden, Keith, 158–59, 160
Wallauer, Claiton, 181
walnuts, 2–3, 4, 13, 23, 30, 100, 223; origins
of, 8; poisonous effect of, 89–90
Wangenheim, Julius von, 6
Washington, George, 27–28, 30
water, 66–71, 75–76, 95, 97, 114, 116, 124,
190; desalination of, 191–92. *See also*
irrigation
Went, Fritz, 85
Weston, Richard, 5
West Texas Pecan Nursery, 45
Wight, J. B., 109
Wilson, E. H., 215
Wolstenholme, Nigel, 188
Woodson, Robert, 140

Yguase Indians, xiv

zinc, 77, 80–81, 113–14
Zuni Indians, 21, 22